25

ESSENTIAL SKILLS

& Strategies *for the* Professional *Behavior Analyst*

25

ESSENTIAL SKILLS

& Strategies *for the* Professional *Behavior Analyst*

Expert Tips for Maximizing Consulting Effectiveness

by Jon Bailey & Mary Burch

Routledge
Taylor & Francis Group
New York   London

Routledge
Taylor & Francis Group
270 Madison Avenue
New York, NY 10016

Routledge
Taylor & Francis Group
27 Church Road
Hove, East Sussex BN3 2FA

Printed in the United States of America on acid-free paper
10 9 8 7 6 5 4 3 2 1

International Standard Book Number: 978-0-415-80067-9 (Hardback) 978-0-415-80068-6 (Paperback)

Library of Congress Cataloging-in-Publication Data

Bailey, Jon S.
    25 essential skills & strategies for the professional behavior analyst : expert tips for maximizing consulting effectiveness / by Jon S. Bailey & Mary R. Burch.
        p. cm.
    Includes bibliographical references and index.
    ISBN 978-0-415-80067-9 (hardback : alk. paper) -- ISBN 978-0-415-80068-6 (pbk. : alk. paper)
    1. Behavior analysts--Professional ethics. 2. Behavioral assessment--Methodology. I. Burch, Mary R. II. Title. III. Title: Twenty five essential skills and strategies for professional behavior analysts.

RC473.B43B35 2009
616.89--dc22
                                                                    2009010305

Visit the Taylor & Francis Web site at
http://www.taylorandfrancis.com

and the Routledge Web site at
http://www.routledgementalhealth.com

Dedicated to W. Scott Wood, PhD,
who spent his entire academic career
modeling professional behaviors for the rest
of us. He was a superb colleague, master
teacher, and elegant researcher. He was
our friend, and we miss him terribly.

Contents

Acknowledgments

We would like to thank Amanda Prestemon for the many hours she spent reviewing this manuscript. With nearly 10 years' experience in applied behavior analysis, Amanda was able to identify gaps in content and places where additional explanations were needed. As a graduate student in a master's Applied Behavior Analysis program, Amanda also provided excellent feedback regarding whether new behavior analysts would be likely to follow the specific advice offered in the chapters. Maxin Reiss, Mary Riordan, and Ken Wagner all provided important input on real-world behavioral consulting. Countless others submitted scenarios and participated in interviews that were enlightening and informative. Graduate students in the Florida State University PhD and master's programs educated the first author for nearly 40 years on the potential our field has to do good in our culture. They were the inspiration for writing this book. Aubrey Daniels' contribution to the field of consulting in performance management is historic, and his influence is immense. Aubrey is a role model of perfect professional demeanor at all times, and he sets a standard of productivity that is unmatched. He is a modern-day hero.

Finally, we would also like to thank our editor, George Zimmar, who believed in this book from the very beginning and could see the need for a companion work to our *Ethics for Behavior Analysts* book.

Preface

I come home from working at the Developmental Disabilities Center twice a week, drop on the couch, and just cry. I don't know what's wrong with me. I think they just don't like me and don't trust me. I feel like an outsider. I have clients whom I love, and I enjoy the challenge of solving problems. I'm well paid by my consulting firm, but at the DD center they don't respect me, and they won't listen to me. I've been told the administrator talks about me behind my back. They like to use drugs for treatment instead of my behavior plans. … I can't admit to my supervisor that I'm in trouble. I don't know what to do, really, I don't. I'm board certified, and I've taken Dr. Bailey's ethics course, but it's not helping me in this situation.

This emotional and heart-wrenching plea came to us in the form of a desperate phone call from Kimberly, a newly certified behavior analyst. This extremely bright, enthusiastic, go-getting graduate student had such an intense desire to get her first job and begin helping clients with behavioral needs that no one would have predicted she would find herself in the depressing situation she described. But she did.

We began to notice that many other behavior analysts were experiencing similar problems, and we had a revelation—being an expert in behavior analysis is *not* sufficient for a behavior analyst to be a successful consultant. As our field continues to grow, it is critical that we educate behavior analysts on *all* of the skills needed to be effective and make a difference in the life of others.

Applied behavior analysis evolved from the experimental analysis of behavior in the mid-1960s. Our field became formalized in 1968 with the publication of the first issue of the *Journal of Applied Behavior Analysis* under the editorship of Mont Wolf at the University of Kansas. The blueprint for the field was established in a seminal article in that issue, "Some Current Dimensions of Applied Behavior Analysis," authored by Don Baer, Mont Wolf, and Todd Risley (1968). In this article, they outlined the key distinctions of this new field that made it different from the rest of psychology. As described in the article, behavior analysts were interested in solving applied problems by using a to-be-developed technology based on the science of behavior; that is, operant conditioning. This technology would be inherently data based. It would include its own methodology for demonstrating cause–effect, that is, the single-subject research design, and it would evolve over time to give us a vast array of techniques that would show how these procedures would help people improve the quality of their life. In 1968, the cutting-edge thought-leaders who authored this important article did not anticipate today's overwhelming demand for behavior analysis. This enormous need for services has developed in the past 5 years, and services are now provided in many countries by Board Certified Behavior Analysts®.

As a result, master's degree programs have sprouted like wildflowers across the United States and indeed the world. Two-year and three-year graduate programs that turn out behavior analysts by the hundreds are now working overtime to provide professionals to work with individual clients who are autistic, developmentally delayed, brain injured, or otherwise disabled. In some cases, behavior analysts are working one-on-one with clients, and in other situations, they are working with teams of paraprofessionals who are implementing behavior programs designed by a behavior analyst.

Behavior analysts are also working in business, industry, government, and organizational settings to improve human performance in safety-related areas or to increase productivity, product quality, or service. In these settings, the behavior analyst takes on the role of the consultant, the professional advice giver who must know a great deal about how organizations work and don't work and about how to train, motivate, and manage people in settings that were never designed from the outset to be optimal for human performance.

As it turns out, being an expert in behavior analysis does not provide all of the necessary skills to be an effective, successful consultant. The settings where we work have often been visited before by other consultants who had no behavioral training whatsoever but who, with their finely tuned sense of business etiquette, social skills, and gift of gab, have made it difficult for the behavior-technology-savvy behavior analyst to make much headway. Upper-level management of human-service organizations and CEOs of major corporations now have an expectation of a quality of interaction that is hard to acquire in graduate programs that offer only courses in applied operant conditioning, research methodology, functional analysis, data collection, and practicum experience working one-on-one with an autistic child. And it turns out that working as a consultant in a developmental training center, in a classroom for children with behavior disorders, or with parents who need to learn how to manage their unruly children requires that the behavior analyst must interface with a wide variety of people who present (a medical term meaning "show up with") an amazing array of contentious and obstructive behaviors that can thwart the unwary and unprepared would-be behavioral consultant.

This became obvious when the first author was contacted by the supervisor of a recent graduate—a hardworking and bright individual who was failing on one of his first consulting assignments. This budding behavior analyst was yet another young professional who found himself in a dilemma much like the one

Kimberly described to us (in the case at the beginning of this preface). According to the supervisor, the new behavior analyst had missed the initial cues from management that he was in trouble, and when he finally learned there were problems, he did not seek help. His tendency was to blame the direct-care staff for their shortcomings and failure to carry out his programs. Upon further investigation, he had simply been unprepared for a semihostile school environment that paid lip service to wanting behavioral consultation but in truth was set in its traditional ways.

Rather than rebuke or blame the new consultant, the first author made an attempt to determine what went wrong in his training. This lead to countless interviews with current and former students, supervisors of consultants, trainers of consultants, and CEOs of companies that hired behavior analysts. In addition, senior consultants were asked a series of questions about their experiences dealing with tough problems in a variety of settings, how they solved these problems, and what they had learned from the experience. When possible, these consultants and supervisors were asked to provide working scenarios that described in a concise format the nature of the problems encountered.

From these interviews and written scenarios, we developed over a 6-month period key words and descriptors of skills and strategies. At this stage, about 100 descriptive terms emerged as important skills and strategies for successful behavior analysts. This number of skills was clearly too many to try to describe or teach. We began a search of sources of knowledge that might prepare the consultant in training for the difficult road ahead. Using key words and Amazon.com, we found it was possible to determine recent books that seemed to focus on the key skills we identified, even though they were not specifically written for our new type of professional: the behavior analyst consultant. Although these books were written for professionals in other fields, it was clear that they had relevance for us. They denoted and described general categories that are clearly required for any professional operating in someone else's setting; topics such as business etiquette, assertiveness, and leadership were common. We found that the business consulting literature emphasized that professionals should have excellent personal communication and persuasion skills as well as a strong background in negotiation, lobbying, and public speaking. As categories of skills and strategies that began to encompass our original list of 100 emerged, a solution to the problem of categorizing our comprehensive list of required skills began to look possible. By reanalyzing the scenarios that we had gathered in terms of "What skills would it take to fix this?" we eventually were able to formulate five general categories of skills and strategies. In addition to having the basic skills just described, consultants would have to be prepared to apply their knowledge of behavior analysis to deal with what is known in the business literature as "difficult people." The good behavior analyst has to know how to use his or her knowledge of functional analysis, shaping, and performance management to deal with these problems and ask pointed questions about issues that come up daily with "Can I see that?"

As a professional, the behavior analyst must also confront the difficult task of managing his or her own behavior on a daily basis. Without careful monitoring, even a bright, highly motivated behavioral consultant can waste time, be a burden to other professionals, get stressed out, and find him- or herself needing help but not knowing how to go about getting it.

One final area emerged from our interviews and scenario collection. There is an expectation that behavioral consultants will grow over time, roughly a 5- to 7-year period when they are expected to take senior consultant positions. These positions will involve additional responsibility, and the need for greater wisdom in making decisions can have a wide-ranging impact on the organizations served. From preschools to factories, experience teaches consultants to refine their critical-thinking skills and to anticipate and quickly troubleshoot problems that invariably arise while consulting in any setting. Senior consultants are expected to take on training, coaching, and mentoring roles with newly minted behavior analysts and may engage in these important tasks with mid- and upper-level managers as well.

Finally, with time and experience, advanced consultants are expected to begin to see the "big picture" of how the world works and to develop an appreciation for the larger metacontingencies that control our society and our nation. This big picture analysis then expands to a broader worldview in which the consultant can suddenly begin to see the behavioral connections between his or her failed efforts to persuade a school principal to adopt a new discipline policy and the failure of an emergency relief effort in Myanmar.

The consultant who has developed advanced skills will have developed one of the most important skills of all—aggressive curiosity. Aggressive curiosity is the skill and attitude about the science of behavior that will enable the advanced consultant to see the beauty in measurement techniques that are robust enough to document the behavior problems of a client with Prader–Willi syndrome who routinely goes AWOL, track the cell phone usage of people in third-world countries, or monitor the feeding patterns of Antarctic penguins.

For the modern-day behavior analysis consultant, being competent and well trained in the technical aspects of behavior analysis is simply not enough. To be successful and effective, behavior analysis consultants need knowledge in critical areas of competence, which now also include essential business skills, basic consulting repertoire, the ability to apply behavioral knowledge, vital work habits, and advanced consulting skills.

25 Essential Skills & Strategies for Professional Behavior Analysts was designed to be used as a companion to our book *Ethics for Behavior Analysts* (Lawrence Erlbaum Associates, 2005), in courses addressing ethics and professional issues in behavior analysis, or as a handbook for practicum courses where students are acquiring and testing their consulting skills for the first time. Supervisors of newly hired behavior analysts who consult in school systems, in residential facilities, or with families should also find the taxonomy useful in spelling out what their expectations are for professional representation of the consulting firm. Finally, experienced consultants might find the references to the professional consulting literature and checklists of value in improving their own skills.

Behavioral consulting is largely the art of practicing the science of human behavior. We hope that this book conveys the excitement and challenges that face our new colleagues as they join our ranks as professional behavior analysts.

Jon Bailey and Mary Burch

One

Essential Business Skills

1

Business Etiquette

The CEO of one large consulting firm said,

> *They have to fit in. I honestly don't care if they have visible tattoos when they're with their friends, but tattoos and facial hardware won't fly in our full corporate environment where a business suit is considered the uniform for men. I abruptly ended an interview with one young man when he showed up with a ring in his eyebrow. We charge nearly $5,000 per day; our clients have certain professional expectations. Our consultants have to know how to put people at ease. It doesn't matter how smart you are if you can't make a good first impression and sell our services to a potential business partner.*

In the past 40 years, our field has evolved into a full-fledged profession. If we look objectively at ourselves as a profession, we would have to say that we have a "high-touch, high-tech" service to deliver and a rather unique one at that:* *Behavior Change on Demand* is the way one luminary at a national conference recently described our potential. In many respects, what we have to offer is comparable to other premium professional services such as

---

* *High touch* means that our service is very personal and individualized to each client. Of course sometimes we actually do touch our clients when we do graduated guidance or apply some procedures, such as time-out. *High tech* means that we employ a science-based technology of behavior; procedures are based on highly refined procedures that are strictly controlled by protocols that have been vetted by experts in the field.

3

top-flight legal advice and representation, first-rate medical and dental care, and four-star IT support from a premier dot-com software company. We are not yet charging the same rate as attorneys, but in some parts of the country, there are comparable billable rates for master's-level behavior analysts, MD psychiatrists, and PhD clinical psychologists.

OUR CURRENT IMAGE

We have a long way to go in terms of our image. Today, if applied behavior analysis is compared directly to other premium professional services in terms of how we present ourselves to our consumers, most people would see quite a contrast. Our "service representatives," the Board Certified Behavior Analysts®, are often young people with a casual attitude and even more casual dress. Many address each other on a first name basis, and this casual demeanor extends to the way they present themselves to other professionals.

Behavior analysts are in a rather intense competition with people in other professional services, and even though this has been the case for quite some time now, most behavioral consultants do not seem to be aware of this development. The greatest pressure seems to be, somewhat surprisingly, in the area of autism treatment. This

> "No other treatment currently available has the breadth and depth of applied research showing clinically significant changes in behavior."

is somewhat difficult to believe, because from an evidentiary perspective there is essentially no competition. No other treatment currently available has the breadth and depth of applied research showing clinically significant changes in behavior. Unfortunately, the image presented to the public—our potential consumers—by many representatives of our field is that of a gaggle of behavioral geeks who spout technical terms that sound somewhat ominous and threatening. *Control, reversal design, contingencies, manipulation,* and *intervention* do not sound particularly user friendly. Furthermore, when used to describe how they might be used in treating a child with autism, these expressions can be downright scary. A child with autism is somebody's baby, son, little boy, little girl, or daughter. What parent wants to hear a professional say, "We're going to extinguish Bradley's behavior." Loving parents want us to help Bradley, not extinguish him.

Compare this with the smooth and soothing talk of the well-prepared, perfectly groomed, more mature and seasoned competition, and you will see quite a contrast. To our credit, the next generation of behavior analysts is more highly trained and better prepared than any previous group. These behavior analysts are enthusiastic and technically skilled and have a clear focus on results, and they have the drive and tenacity to stick with a child. But as we'll repeat a number of times throughout this book, being technically competent alone is no longer enough for the behavior analysis *professional.*

First impressions do count, and the very first of the 25 professional skills that needs to be in place is business etiquette. Business etiquette skills range from arriving on time for every appointment to debriefing the administrator before you leave the premises. It often means sending a handwritten "thank-you" card. Let's look at the specifics that make up contemporary business etiquette.

FIRST IMPRESSIONS COUNT

Ours is a very "high-touch, high-tech" service where we get very close not only to our clients but to their families, their physicians, and teachers, so there are a lot of first impressions that have to come off perfectly. The following is an example of one *first impression* that really needs to work: "Dr. Samuels, this is Becky's behavior analyst. I asked her to come with me today because we want to evaluate Becky's medications to see if they are helping her in school." The behavior analyst can't look like she just came from the playground if she is going to be determining if the physician's prescriptions are effective in improving Becky's on-task behavior. Or, consider an introduction to the principal at a school by the behavior analyst's supervisor: "Good morning, Ms. Hanover. I'd like to introduce you to my new behavior analyst, who will be working with you this year. Gretchen got her master's in applied behavior analysis from Caldwell College, and she also has an undergraduate degree in special education with a minor in child development." If Gretchen is standing there chewing gum with her midriff showing, we can tell you there is going to be a problem with her credibility on the IEP (individualized education program) committee. Although behavior analysts don't have anything close to a uniform set of guidelines for personal appearance, there are some general suggestions that will prevent embarrassment.

I SEE LONDON, I SEE FRANCE

Dress in a manner that is acceptable to leaders or supervisors in the setting where you will be working. In the corporate setting for performance management behavior analysts, business suits and dresses will most likely be the standard attire. It is usually clear what the dress code for the corporate job will be.

For those behavior analysts who work in schools and therapeutic settings, make sure you are not violating the dress code for the agency in which you are working. For example, some schools prohibit staff from wearing open-toe shoes (sandals). Know the rules with regard to the dress code in the settings where you are providing services. Although we generally say "no jeans," there may be a few situations in which a behavior analyst is expected to be in a sandbox or to roll around in the dirt. In these cases, if you'll be attending a meeting later in the day, you should carry a change of clothes in your car. Don't show up to a meeting and expect everyone to understand why you are dressed for the playground.

Acceptable attire for women in the day-to-day job as a behavior analyst includes what is generally referred to as "business casual" attire: oxford cloth shirts, dressy knit tops, scarves, blouses, vests, fitted or dress sweaters, and blazers or sport coats with business casual khakis or slacks and business casual skirts and dresses. If your consulting firm has a polo-style shirt with the company logo, this is fine too.

Examples of unacceptable attire include suggestive, risqué, or revealing attire; clothing made of sheer or see-through fabrics; sweatshirts and T-shirts; athletic wear; stiletto heels; oversized sweaters; sundresses; crop tops; midriff tops; tube tops; tank tops; sleeveless tops; undershirts; flannels; miniskirts; halter tops and halter dresses; sweatpants; jeans; leggings; stretch or stirrup pants; convertible slacks and shorts; cargo pocket pants and slacks; tight-fitting pants (we don't need to know if you are wearing a thong); spandex or Lycra anything; nylon jogging suits; novelty buttons; baseball-style hats; gaudy jewelry; and similar items of casual attire that do not present a businesslike appearance.* You might have a tendency to think jeans and T-shirts are OK in settings such as preschool classrooms "because that is what the staff wears," but remember that you are trying to establish yourself as a professional, and "student clothes" will not help you gain the respect and credibility you need.

For men, oxford cloth shirts, clean and pressed polo shirts, and blazers or sport coats with business casual khakis or slacks will usually be acceptable. Not acceptable are jeans, T-shirts, tennis shoes, and baggy pants with underwear showing. For men it is expected that any facial hair will be neatly trimmed. For both sexes a generally conservative appearance is strongly recommended, with no tattoos showing and no facial hardware visible.

For meetings and presentations, you'll want to step up your level of dress. For men, a jacket is recommended for important

* This list was compiled from several human resources Web sites for retail and professional settings.

meetings, such as when meeting with the principal or a physician. For women who are attending meetings where they need to have a presence, traditional business clothes should be worn rather than the daily company polo shirt and slacks that are worn when working with children.

GETTING TO KNOW YOU

If you are meeting someone for the first time, it is a good idea to do your homework so you know something about the person. Saying, "I'm delighted to meet you, Ms. Hanover. I understand that your school was ranked in the top 10% in the district last year" is much better than saying, "Nice to meet you." It goes without saying that being a little early for meetings is a good idea (10 minutes is recommended) and that being late is never acceptable. If you arrive 10 minutes early, you will have time to visit the restroom to check your appearance and your breath one last time and to practice your smile.

Some standard recommendations for that *best* first impression include making good, but not scary, eye contact and presenting a firm handshake; the deliberately limp four-finger handshake and the "air kiss" are definitely out except in certain neighborhoods in Manhattan. Also out is the use of slang expressions, curse words, and substitute curse words such as "bleepin' " or "friggin'."

Make sure that you use the person's proper title and use his or her name in the meeting. If you have a tendency to forget names, you better develop a mnemonic for this or make quick notes when no one is looking. When there are several important people at a meeting, make note of everyone in attendance. Another strategy is to ask people, "Do you have a card?" as you hand them yours.* File these cards away, and enter the information on your

---

* If your company does not provide you with business cards, get them for yourself. You can find perforated card stock paper in most office supply stores that can be used in your own computer printer. Business cards don't have to be elaborate; your name, degree, Behavior Analyst Certification Board designation, phone number, and e-mail address are sufficient.

computer database so that you can quickly refer to your notes for future meetings.

MEETING ETIQUETTE

Many meetings are short, stand-up affairs, sometimes in a hallway or in the outer office area in a school or corporate headquarters. But if you are going into someone's office and are going to be seated, pay attention to the seating pattern. Assuming you are the visitor, pause briefly to see if your host signals where you are to sit. A good host will pick up on this cue and put you where he or she wants you seated. If it seems no one cares where anyone sits, place yourself where you will have good eye contact with the key players in the meeting. Make yourself comfortable, and be ready to take notes.

If you don't know the host of the meeting, as soon as you enter the room, smile, say hello, and introduce yourself. If you need to introduce someone you've brought to the meeting (such as your supervisor), make sure you are good at introductions. Be ready with not only the names of the individuals but also their job titles and a few brief words about what they do.

In cases where beverages and food seem to have been ordered primarily for your benefit, accept something to eat or drink when offered. It is discourteous to refuse food and drinks when your hosts have gone to a lot of trouble. You don't have to drink or eat a lot, but you should appear appreciative of the gesture. Don't make a big deal about your special requirements for your food or beverages. Now is not the time to point out that you think of yourself as a coffee gourmet and prefer to drink coffee from a specific $6.00-per-cup franchise or that you take only 2% milk in your coffee and nothing else will suffice.

As Dale Carnegie pointed out more than 70 years ago (Carnegie, 1936), the one thing that people enjoy the most is talking about themselves. You will be a very popular person if you focus your attention on the people you are meeting. Asking them questions based on what you see in their office is a great way to show that you care about them and that you are observant. Saying, "What beautiful children, what grade are they in?" or "Is that your dog? Is it a Maltese? I love dogs" tells people that you noticed the photos on their desk. If possible, in the first meeting see if you can find something in common with the person you are meeting, for example, you enjoy the same sports, root for the same team, went to the same school, and so on. Remember, however, to keep the icebreaker conversations short and quickly get down to business.

BUSINESS MEALS

We're sorry to disappoint food and wine aficionados, but behavior analysts aren't at the top of the list of the most finely wined and dined professionals. Many behavior analysts who work with special needs clients feel grateful if they have a few minutes to grab a burger as they rush from one school to the next on a busy day. Consultants who work with businesses are more likely to have their presence requested at business lunches and dinners.

The main thing to know with regard to business meals is that in some cases, the negative impression caused by bad table manners can actually result in a consultant not getting hired, a contract not getting signed, or a behavior analyst not getting a job. There are numerous books and Web sites on manners.* If you are going to be attending business meals, get one of these books and read about dining etiquette specifics, such as how much to tip, which utensils to use and when, how to choose wine, where to place your napkin, how to make a toast, and which food items to avoid (hint: watch out for that spinach on your teeth and finger foods that are a mess).

WHEN IN ROME

Behavior analysts are finding jobs and attending conferences throughout the world. If you plan on working as a behavior analyst in another country, get a book on corporate manners for the

---

* Go to www.ravenwerks.com/practices/etiquette.htm for a sample.

country that you'll be visiting, or talk to someone who can advise you about the country's manners and cultural differences. In the United States, we encourage behavior analysts to give direct feedback. In some countries, a person who gave direct feedback to others would be viewed as rude or aggressive. If you will be working in another country, be sensitive to cultural differences, and learn what you need to know to give behavior analysis a good name around the world.

CELL PHONES AND TEXT MESSAGING MANNERS

As technology continues to evolve, so do the manners related to the use of new equipment. It wasn't long ago that only a few people had cell phones, and those phones were nearly as big as a shoe. Now, it seems everyone in the United States older than the age of 10 has a cell phone. Correction—we've seen 6- and 7-year-olds in restaurants with cell phones. One parent told us she wanted her 6-year-old son to be able to call her in case of an emergency. She said she felt great peace of mind knowing that if her child was kidnapped, he could be tracked because of his cell phone (assuming, we guessed, that the kidnapper was dumb enough to let him keep his phone). So cell phones are here to stay, and future behavior analysts are learning to text message their friends before they can spell.

When you are working with a client, your phone should be turned off (not on vibrate—OFF). People can leave a message, and you can return the call as soon as the session is over. When you are in a meeting, your phone should be turned off; you don't need a cold, icy stare from an administrator because *your* cell phone interrupted her meeting.

When you are providing services, the client and/or agency should have your full attention. Excusing yourself to go out in the hall to take a call from your mother or boyfriend is not acceptable. The same rules apply for text messaging. When you are at work, your job, the clients, and the professionals with whom you are working should have your undivided attention.

Confidentiality applies when you are talking on a cell phone— don't talk about your agency, staff, or clients when others are around. This means don't go to lunch or sit in the waiting room at a doctor's office and talk on your cell phone about a client. If anyone can hear what you are saying, you've violated confidentiality.

News flash for avid cell phone users: People in restaurants, on public transportation, and in waiting rooms don't want to hear your conversations. We simply aren't interested. The first author routinely shapes on the behavior of cell phone users who are engaging in noise pollution. He walks up to them and nicely says, "Could you please take your call outside? We don't want to hear about Frank's car problems." They usually look horrified that someone actually heard what was said, and they quickly scurry outside with their phone.

We aren't alone in promoting good manners for cell phone users. Theaters attempt to gain some antecedent control by showing a reminder on the screen before the movie begins that tells the audience to silence their cell phones and take their crying babies to the lobby. Furthermore, the month of July was designated National Cell Phone Courtesy Month. As an increasing number of public places are regulating how and when cell phones can be used, cell phone users are encouraged by the communications industry to use their phones responsibly and with consideration for others to prevent further restrictions on usage.

IMPRESSION MANAGEMENT

In addition to having good manners and showing consideration for others by dressing professionally, showing up on time, and not interrupting a meeting with your personal cell phone calls, it is important to enter each *first* meeting with the goal of leaving the person with a certain impression of you. When possible, do your homework, and find out in advance about the person you're meeting. Making a comment such as "I heard that you are the administrator who was responsible for starting the NewStart Program" can help you get to know the person. If you want parents to know how much you care about their child, you should say this explicitly. Give specific examples of how you care, and perhaps tell a story about yourself that makes it clear you care about children and their child in particular, being careful not to suggest an inappropriately close attachment. If your goal is to leave the physician with the impression that you are competent in evaluating medications for children, you can give an example of something you've done recently that demonstrates this. For example, you might say that you just completed a study at Monroe Middle School with five children who were on Ritalin. When you are dealing with professionals who have limited time to meet with you, do not let yourself get drawn in to some lengthy, unnecessary conversation about a sitcom, pop star, or recent debate between political candidates. Whether you are with parents of children with attention deficit/hyperactivity disorder, teachers of students who are developmentally disabled, fire safety supervisors, process-analysis leaders, or corporate executives, your goal in meeting people is to create the impression that you are a serious, highly qualified professional. Good business etiquette skills will help you convey your message: You are a well-trained, competent behavior analyst; you *can* make a difference; and you're here to help.

SUMMARY

Behavior analysts complete a rigorous course of study to acquire the knowledge and skills needed to understand and change behavior. But for behavior analysts to become well-respected professionals, technical competence alone is not enough. Behavior analysts who want to be successful as consultants must develop skill sets in the general arena of business manners. Dressing for success, using down-to-earth language that clients can understand, knowing how to introduce oneself, and having proper meeting etiquette and manners related to cell phone use are practical work-related behaviors that can help the behavior analyst who has a desire to be a successful, professional consultant.

FOR FURTHER READING

Bixler, S., & Dugan, L. S. (2001). *5 steps to professional presence.* Avon, MA: Adams Media.

Carnegie, D. (1936). *How to win friends and influence people.* New York: Pocket Books.

Detz, J. (2000). *It's not what you say, it's how you say it.* New York: St. Martin's Griffin.

McIntyre, M. G. (2005). *Secrets to winning at office politics: How to achieve your goals and increase your influence at work.* New York: St. Martin's Griffin.

Whitmore, J. (2005). *Business class: Etiquette essentials for success at work.* New York: St. Martin's Press.

2

Assertiveness

Without mentioning any names, a program director described one applicant:

> *She looked fabulous on paper, her grades were amazing, she was at the top of her class, and on her first attempt, she passed the BCBA [Board Certified Behavior Analyst®] exam with a high score. We were all looking forward to interviewing her because she was everything we needed. But when she came to the interview, she was a mouse. It started with a limp handshake and went downhill from there; we could barely hear her when she talked. One question I always ask is, "Can you describe a situation where you had to show some assertiveness to accomplish a goal for a client?" She paused, and then paused some more and said, "I usually just show some data, the data speak for themselves." After she left, it was the consensus of our team that she would have an extremely hard time managing our staff or gaining their respect.*

In therapeutic and educational settings where behavior analysis services are provided, there are many situations in which a treatment under consideration is not in the best interest of an individual client. Behavior analysts often find themselves in the position of being the professional on the treatment team who needs to speak up for the client and advocate quality behavioral programming.

Sometimes a supervisor or colleague may ask the behavior analyst to implement a procedure that is unwise or unjustified based on the status of the client or the environmental circumstances. As behavioral scientists, we have been trained to analyze behavior, ask the right questions, take data, find the functions, and present our findings in a logical fashion so that everyone can make a decision that is best for the client. This approach is a "best practice" approach, and most often the end result is that treatment team members will support the systematic implementation of the behavior plan all the way from the functional assessment to follow-up data. But occasionally there are professionals on the treatment team who will attempt to resist or reject the behavioral approach or the suggestions of the behavior analyst. When resistance and "push back" occur, the behavior analyst will have to consider being more assertive to influence the decisions that are about to be made regarding a client's treatment.

> "Being assertive requires an in-depth knowledge of the people with whom you are dealing, and it is appropriate only under specific circumstances. Knowing how and when to be assertive is a critical skill for effective behavior analysts."

Using an assertive approach is difficult for behavior analysts who don't have experience in playing what is commonly referred to in the corporate world as "hardball." Assertiveness is risky. Being assertive requires an in-depth knowledge of the people with whom you are dealing, and it is appropriate only under specific circumstances. Knowing how and when to be assertive is a critical skill for effective behavior analysts.

ASSERTIVENESS ON BEHALF OF YOUR CLIENT

One of the most frequent occasions for you to be assertive is when a group decision is to be made and you have input as the behavior analyst. In team meetings, there will be many times when someone in the group will be pushing for action that in your opinion is not warranted. As you listen closely, you might determine that because this person has taken a very strong position and has presented it in an adamant, emotional manner, everyone is inclined to just go along. People often go along with a bad plan simply because it is the easiest thing to do. Furthermore, there will be some people at the table who are motivated by thoughts such as, "Why rock the boat and slow everything down?" "If we hurry we can go to lunch sooner," or "I don't want to make this woman mad; I have to work with her at several schools." As an ethical behavior analyst, before you decide it is time to be assertive on an issue, *make sure that you are right* and that your proposal to the group will improve the client's well-being. Remember, to be effective you can't be assertive on *every* single issue, only on a few. Make sure this is *the issue* for which you are willing to push.

ASSERTIVE BEHAVIORS IN MEETINGS

When you are ready to assert yourself in a meeting, you'll have to do several things simultaneously. First, sit up straight, put your hands in front of you on the table, and calmly fold them. When it is your turn, present your information in a logical, organized, and concise fashion. When you do this, people are likely to agree with you. But sometimes there will be a professional who does not agree and, furthermore, who thinks her approach is the only one to consider. This is when you will need to use your very best assertiveness skills.

Make good, strong eye contact with the person who is bullying the group into her way of thinking. In your own words, say something that basically conveys the following message: "With all due respect to Sue, I have to say I disagree with what is being proposed here. I'm concerned that [client name] is not being well served by what we are about to do. Here are some problems with this approach ..."

If you aren't successful after explaining why you think the recommended approach presents problems, you may have to be a little more assertive: "I can't agree with our moving forward with this. It appears to me that what you are about to decide is to use the solution requiring the least effort. The easy way is not the right way in this case. Let's slow this process down and reconsider our options."

As you speak, move your gaze from one person to the next at the table. Do not raise your voice, do not go shrill, and don't squint, roll your eyes, emit loud groans of exasperation, or make faces. By all means do *not* apologize for what you are saying. Then re-present your strategy for the solution, the right solution, to the group. Start by using your differential reinforcement of other behaviors (DRO) skills to recognize those parts of the decision with which you do agree. Emphasize how your plan benefits your client; stick to this message, and do not go off course and talk about other benefits.

ASSERTIVE AFFECT

To be effective you also need to make sure that in addition to the words you are actually saying, the other aspects of your delivery are just right. Probably the most important thing you can do when being assertive on behalf of your client is to make sure that your body and your voice are in sync and that you are showing strength. Speak with a strong voice, and avoid "up speech" where your sentences end with an inflection upward and a question mark. If you are a younger behavior analyst, save the common, current young person vernacular for when you are with your friends. Saying, "So,

> "Probably the most important thing you can do when being assertive on behalf of your client is to make sure that your body and your voice are in sync and that you are showing strength."

my client, he was like, really like having a tantrum? Like he was like all upset?" is not appropriate in a meeting.

You can't slouch in your chair and be assertive, and you can't sit up straight and look down at your paperwork and be assertive. Your strong voice, not whiny or angry but strong, steady, and controlled, is your primary weapon in being assertive.

You will receive questions from the meeting participants. Answer questions directly, with the same steady gaze and steady voice as when you presented your professional opinion. Give direct answers, and do not let people take you off message—they *will* try to do this. Do not reinforce comments such as, "You're just trying to have your way on this." You should follow questions with a pause, then go back to the proposal. Make it clear that your proposal is straightforward. Keep describing ways that the approach you are suggesting will improve the client's life and how it can be evaluated for effectiveness. Offer to abandon the strategy if you are proven wrong. After all, you are a person who relies on data and who in this case just happens to also be advocating for the client. Afterward some people will comment, "I didn't know you had it in you." Or "That was amazing. I had no idea you felt so strongly about that decision." When the day comes that you receive comments such as these, you'll find that your resolve is strengthened, and you are more committed than ever to being a good advocate for your client.

ASSERTIVENESS ON YOUR OWN BEHALF

The second major way in which you will need to be assertive is on your own behalf. This is most important when you are new on a job and requests are flying at you left and right to do this and do that and by the way, hurry up, we needed it yesterday.

The initial 90 days on a new job most often come with an unwritten grace period where you can ask questions that will later appear to be in the category of "stupid questions" but in the beginning are considered appropriate for a newbie. So your first act of assertiveness on your new job will be to ask a lot of questions about how the organization works and exactly what is expected of you. If someone asks you to do something that you are not sure of, begin by being assertive. Say, "I don't understand," and then rephrase what has been requested to see if this results in a better description. When you are asking questions about a request that has been made of you, you'll want to have a very steady tone to your voice. Don't sound whiny or like you are trying to get out of something; you don't, you just want to understand what is expected. The first 90 days are when you can ask about the table of the organization so you can understand what the power structure is within the organization. If the nurse is the primary decision maker, you need to know this. After 90 days you will need to use your best autoclitics: "I'm sorry, I know I should know how this report should be formatted, but somehow I misplaced that file you gave me."

A new behavior analyst told us about an excellent example of a situation that could have been avoided with some job clarification on the front end:

> A lot of the challenging behavior exhibited by the clients on my caseload is maintained by pain or discomfort associated with illness, injury, or medication side effects. However, many nurses with whom I work are not on the same page as I am. I feel that I need to rule out medical variables that may be related to the behavior before implementing restrictive behavioral procedures. I am often told by the nurse that Client A "just does that" or "he is just acting the fool." I think that Client A may be hitting himself in the cheek because he has a decayed tooth that needs dental attention; he has a history of dental problems and behaviors that are associated with them. The nurse thinks that he is acting up because of the full moon. She is the one who would be allowed to call the doctor to see if some pain medication or other medical intervention might be given until a dentist can be seen, but instead she just ignores my suggestions. She makes it clear that she wants me to issue wrist restraints to deal with the problem. It's a weekend, and I can't take this situation to

anybody higher up the ladder. This is typical for how it goes for me in this job.

This well-trained behavior analyst should have been more assertive in his first 90 days, asking questions about how decisions are made, who he responds to, and how to reach his supervisor in emergency situations. And some assertiveness early on at case management meetings where decisions about client care are made would have been in order. In the case of the client who may very well have severe dental pain, the appropriate assertive response on the spot would be, "No, I can't write an emergency order for wrist restraints; it's not right, and it violates my code of ethics. I think we should call the administrator at home. It's that important."

JUST SAY "NO"

Learning to say "no" is a critical part of assertiveness for any behavior analyst. Our interviews with behavior analysts from around the country indicate that saying "no" is one of the most difficult skills to develop, especially for new, young behavior analysts who are understandably eager to please their new employers.

> "Learning to say 'no' is a critical part of assertiveness for any behavior analyst."

One new behavior analyst we talked with described his frustration at a treatment program for clients with head injuries. It seems that he was led to believe that the orientation of the program was thoroughly behavioral. It soon became clear, however, that the staff wanted him to sign off on "canned" programs that came from a database of programs for previous clients. The behavior analyst wanted to do individualized functional assessments but was told, "We already know the cause of the behavior. Just plug in his name, and sign at the bottom." When he used his assertiveness skills and said "no," the administration fired him during his probation period without cause. About 6 months later, the facility closed after an investigation pertaining to their billing practices and data falsification. Being assertive on the front end of your employment, even during the interview phase, can save you from finding yourself in a job that was not what you bargained for or, worse yet, beginning your new career by getting fired.

HOW TO SAY "NO"

Paul Simon's classic song describes "50 ways to leave your lover." Although we won't list here all the ways to say "no," we think there are probably at least 50 good ways to say it.

When we suggest that you say "no," of course we don't mean you should literally say "no" and nothing else: "Will you help me with this program for a client?" "No." "We are having a staff meeting— can you present Jason's data?" "No." "Our department is going to have a picnic next month. Can you attend?" "No." Blunt, cryptic responses such as these will have us sending you a greeting card in the mail that says, "It was really nice having you in the field of behavior analysis, until you were fired from your first job."

Please Explain Yourself

Sometimes to say "no" you'll want to *include an explanation*. For example, when someone invites you to a work-related social event, you can decline the invitation politely. This "no" includes an explanation such as, "Thank you so much for inviting me, but I will be out of town that weekend." Or "Thank you so much for inviting me, but I have young children, and I really like to stay home on the weekends."

If you are asked by someone to do something you simply don't have time to do, it is perfectly acceptable to let the person know you are busy: "I'd love to do training for your staff, but I am fully committed with three major projects for the next several months."

If you are good at what you do, people will begin to count on you as someone who can get the job done. Then you may find yourself getting requests to work on projects that are not within your skill set. One way to turn down a request for services you are not trained to provide is to say, "Thanks, John, I really appreciate your support, but I don't think I am the best person to lead this particular project. I really don't have any experience working with adult offenders."

Educating Others

There are times when "no" needs to be accompanied by some words that will *educate people:* As a response to someone on a treatment team saying, "We should use cigarettes and ice cream as the reinforcers for Bob's behavior program because he likes them so much," a behavior analyst might reply, "As much as Bob likes smoking and eating junk food, when used frequently, these are in the category of reinforcers that are harmful, and I don't think this is a good idea."

A much stronger version of "no" is appropriate in situations such as the following: "The review team is coming, and we really don't have all the data we need. I'm sure you can generally remember how the clients did over the past few weeks. Can you fill in these data sheets so we don't get cited?" This "no" should be loud and clear. The function of this "no" is to make other people aware that *you will not participate in unethical or inappropriate conduct:* "No, I cannot do that. It is unethical and probably illegal. I am bound by the Guidelines for Responsible Conduct for my field."

As a treatment team member, you will find many times when you have to say "no" to a roomful of professionals: "With all due respect to Janice, I believe her suggestion that sensory stimulation should be the major focus of treatment is not the way we want to go. There is no data to support this approach as a means of reducing maladaptive behaviors. The first logical step here is to conduct a functional analysis." This is the "no" you will use to *advocate the use of sound behavioral procedures* that are based on science.

How About This Instead?

When you find yourself in the position of having to say "no," sometimes a good strategy is to offer an alternative: "I'm not in a position to take on another school for consulting right now, but I know a very good behavior analyst who might be able to help you." This makes it clear you do want to help even though you are not able to provide the services yourself. If time and current commitments are the issue, you might say, "I am so busy right now, there is no way I can help you with that. My time should free up at the end of the month—if you can call me then, I will be happy to help you."

Saying "No" Is Hard to Do

Remember some of the answers you got when you were a child and asked to do something? "Can we go and get ice cream?" "Maybe later." "Can we go to the lake on Saturday and go swimming?" "We'll see." "Can we get a dog?" "When we aren't so busy." These were all answers given by someone who wanted to be nice and didn't want to come out and say "no." You'll probably hear some of these very answers in the settings where you are providing behavioral services. Behavior analysts should have integrity. Make sure when you say "no" that you're always honest and direct.

ASK FOR WHAT YOU WANT

Another major aspect to assertiveness is *asking for what you want*. If you don't ask, you probably won't get what you need or want, because whoever is making the decisions can't read your mind and doesn't know what is important to you. Requests such as asking for an occasional day off or a chance to go to an annual state or national behavioral conference might be granted, especially if you are highly valued for your creative ideas and you make your requests well in advance.

HOW AM I DOING?

Another form of assertiveness is asking for frequent feedback on how you are doing. Although this might seem like fishing for reinforcers or asking for trouble, regular feedback can help you improve your own performance. Far too often, it is company policy that you will receive an annual performance review. As behaviorists, we know that receiving feedback at the end of a year for something you did 11 months ago is much too delayed to have any effect. Don't be a pest, but asking for feedback at least quarterly is a form of assertiveness that will pay off in the long run. If your supervisor does not want to take the time to put the feedback in writing, you can summarize your meeting and send the notes in an e-mail to your supervisor so there will be a paper trail. By requesting frequent feedback, you can show your supervisors that you have a strong desire to be an excellent employee. This will put you in good stead when you have to occasionally say "no" to a request.

> "Asking for feedback at least quarterly is a form of assertiveness that will pay off in the long run."

SUMMARY

Assertiveness is a key skill for behavior analysts who want to be highly effective advocates for their clients and behavioral services. For behavior analysts working in every type of behavioral setting, knowing when and how to be assertive on behalf of clients and themselves and in meetings is a day-to-day skill. Being able to say "no" in a nice but firm way, asking for what you want and need, and requesting routine feedback are additional assertiveness skills that will lead to success.

The behavior analyst who has well-developed, appropriate assertiveness skills will be a person who can truly make a difference on a treatment team and in the life of a client.

FOR FURTHER READING

Detz, J. (2000). *It's not what you say, it's how you say it.* New York: St. Martin's Griffin.

McQuain, J. (1996). *Power language: Getting the most out of your words.* New York: Houghton Mifflin.

Pachter, B., & Magee, S. (2000). *The power of positive confrontation.* New York: Marlowe.

3

Leadership

The single most important leadership function is to create a focus for the group's behavior.

Aubrey Daniels

The literature pertaining to leadership is enormous. There are more than 1,000 recent books on the market that deal with leadership, and an Amazon.com search for books with the word *leadership* in the title produces an astonishing selection of more than 266,000 books.

Beginning with the writings of the Chinese military general Sun Tzu (544–481 BC) and moving forward to 2008 publications such as *The Breakthrough Imperative: How the Best Managers Get Outstanding Results* (Gottfredson & Schaubert, 2008), there seems to be a consensus about what constitutes a good leader.

Most books tell us that good leaders are intelligent and enthusiastic. They have outstanding social skills, and they lead by example. We're told that good leaders are confident, are not afraid to take risks, and know how to overcome obstacles. Emotional stability is another characteristic frequently listed for good leaders. When the going gets rough and there is a tremendous amount of frustration in the workplace, good leaders are even-keeled. They instill trust, they have integrity, and they empower the people around them. Good leaders have vision, they see the "big picture," and they know how to guide a department or an organization.

Certainly, these general characteristics make for extremely interesting reading, but behavior analysts who want to develop leadership skills know that to be effective, they must translate these traits into observable behaviors. As you are doing behavior analysis consulting, you can adopt some specific behaviors that will help you develop and eventually assume a leadership role.

LEADERSHIP BEHAVIORS: HOW TO GET STARTED

Those who are chosen by an administration to be leaders clearly have the trust of those in the current leadership hierarchy. If the administrators choose you, it means that they have had a good look at your repertoire and believe that you support the

> "To prepare yourself to be a leader in your organization, you need to observe the current leaders in action."

organization, understand its mission, and have the same values they do. To prepare yourself to be a leader in your organization, you need to observe the current leaders in action and determine if your values match theirs. If your values match, then you'll need to explore ways of demonstrating that you can in fact be a leader.

Volunteering for a project that needs to be completed on a short timeline is one good way to test yourself and gain some experience. As a project leader, you'll be guiding others to complete tasks on time and in a coordinated fashion. This kind of activity on your part signals to the present leadership that you have a desire to move up in the organization. Short-term leadership experiences also allow you to learn how to quickly size up colleagues and other volunteers and to determine how best to use their skills. These are all good qualities of a leader.

Working on volunteer projects will also sharpen the skills you need to motivate people. Because your colleagues are not direct reports (you don't supervise them), you face an additional restriction. You won't be able to simply tell them what to do; rather, you must hone your skills of leadership through antecedent control and positive reinforcement. The antecedent control is what the literature on leadership refers to as "vision," or the ability to describe a task or project in such a way that people can "buy in" to produce a desired outcome. This can come in lofty terms, such as Martin Luther King, Jr.'s "I Have a Dream" speech, but the details need to be spelled out as well. In our behavior analysis lingo, you also need to provide the task analysis. Being able to describe how, using the task analysis as the guide, the lofty goal is to be achieved is clearly a key element to leadership.

Additional antecedent control comes in the form of your visible enthusiasm for the project at hand; you can't expect your followers to be excited about a project if you aren't. And the excitement you show also serves as a sample of the behavior that you will engage in as the project moves forward; that is, it shows your group members that you are highly likely to give out generous amounts of praise and approval for their work.

Another characteristic of leaders that is critical to their success is integrity, a commitment to a set of values that is unwavering in the face of pressure from all sides. John Wooden coached UCLA to 10 National Collegiate Athletic Association national championships in 12 years, achieving a distinction that was unmatched in the history of basketball. When asked about his success, he described two sets of rules that his father taught him and his brothers that guided him throughout his 40-year college basketball coaching career: "Never lie; never cheat; never steal. Don't whine; don't complain; don't make excuses" (Wooden & Jamison, 2005, p. 71).

As you work to establish yourself as a leader, you'll need to adopt a code of ethics for yourself that will guide you throughout your professional behavior analyst career. John Wooden was a basketball Hall of Fame coach, an incredibly successful one for four decades at UCLA, and he was clearly a leader. Perhaps it is appropriate to point out that the word *coach* is often used as a metaphor for a type of leadership that can be very successful in settings where the accumulated effort of many people needs to be harnessed to accomplish a particular outcome. *Bureaucratic* or *autocratic* leaders operate out of an older "authority" model that is not very popular these days, but the coach-as-leader model clearly suggests someone with integrity who is guiding a team of people to a successful outcome with a vision, a task analysis, obvious enthusiasm, and generous amounts of contingent reinforcement (Daniels & Daniels, 2005).*

> "The coach-as-leader model clearly suggests someone with integrity who is guiding a team of people to a successful outcome with a vision, a task analysis, obvious enthusiasm, and generous amounts of contingent reinforcement."

VISIBLE LEADERSHIP: PARTICIPATING IN AND RUNNING A MEETING

For the behavior analyst consultant, one of the most frequent opportunities to demonstrate leadership ability will be in the numerous meetings that occur in any organization. As a new employee, you will probably be asked to attend these gatherings where you can observe how the organization runs and how the leadership conducts itself. You won't be asked to do much initially, so observing and taking notes is appropriate. Meetings can provide an opportunity for you to practice your business etiquette (see Chapter 1), show some assertiveness (see Chapter 2), and demonstrate your newly acquired leadership skills.

One tip regarding your behavior in meetings involves understanding the protocol of timing. It is recommended that you always show up a little early; 10 minutes is about right for a new employee. With this standard, you'll probably be the first person

---

* Daniels and Daniels in their book *Measure of a Leader* (2005, pp. 185–190) listed "50 things you can do to increase your leadership impact."

in the room and will have your choice of seating. You'll want to take a seat where you can make eye contact with the chair of the session but not sit so close as to make it appear that you are "sucking up." As people arrive, you can practice your social skills by acknowledging them. Introduce yourself if you don't know who they are; they will appreciate this gesture, and it will put other new people at ease. For example, say, "I'm Jim Harper, the new behavior analyst. I work for Jane on the ESE project that she's starting up." It is appropriate for you to hand out business cards at this time. If they have cards, you can get one from each person. For people who don't have cards, be sure to write down their information (i.e., name, agency, job title, what they do).

When the chair of the meeting arrives, take your seat, and watch what happens as she starts the meeting. Look for good leadership skills: Do people immediately cease their small talk and get down to business? Does she have to "shush" the group to quiet the room? Is there an agenda? (Ideally it should have been shared via e-mail 24 hours beforehand.) Watch to see if the chair specifies a time limit for the meeting (a good sign). Many companies want a record of what went on, and a good chair will assign someone to take minutes of the meeting.

People can be difficult to manage at meetings. They get off task, they go on too long, they are vague in their offerings or suggestions, they don't speak up, and they start bickering among themselves. The responsibility for managing all this falls on the chair. As a new employee, you can note which people present which problems. Good leaders are clear about their objectives for a meeting in terms of both the outcomes and the meeting behaviors: "I have five items for us to discuss today, and I want to be out of here in 1 hour, so please stay on task and help me move this along. Our first item is …" Some big, fast-moving companies such as Google put an image of a large ticking clock on a screen that counts down from the time allotted in order to keep everyone focused.*

---

* See yahoo.businessweek.com/smallbiz/content/sep2006/sb20060927_259688.htm.

LEADERSHIP 101: FIRST STEPS

Assuming that you want to eventually move into a leadership position, learning to participate effectively in a meeting is your first step. Besides being 10 minutes early and introducing yourself to everyone, you'll want to take good notes and look for opportunities to participate in some way. Volunteering to take on a small task where you have relevant skills is a great way to start. This will bring you to the attention of the leadership and put you directly in touch with your colleagues. You will appear to be on board with the organization and selfless in your approach—two good qualities for a new employee. Don't take on a task you can't complete; this will not work out well, and it doesn't bode well for your moving up in the organization.

In addition to volunteering, the next most critical behavior has to do with using the meeting as an opportunity to reinforce your colleagues for their ideas, suggestions, and contributions. If you want people to seek you out later, giving reinforcers is the place to start. Be careful that your praise is not seen as gratuitous, insincere, or immature. Be subtle; sometimes just a well-timed head nod is all it takes to let colleagues know that you appreciate their contribution. If the agenda includes items that you are responsible for, make sure you have done your homework and are ready. Look for the best way to make your points, and be sure to acknowledge anyone else who helped you.

> "If you want people to seek you out later, giving reinforcers is the place to start. Be careful that your praise is not seen as gratuitous, insincere, or immature."

If it looks like you are allotted 15 minutes on the agenda, plan your presentation to be over in 10 minutes to allow time for questions. *Do not run over your time!* At the end of the meeting, confer briefly with the people you will be working with on the project for which you volunteered. You can set up a time to meet and, if appropriate, talk about initial assignments.

As the chair leaves the room, try to catch her to compliment some aspect of the meeting. Very few people do this, and you'll be noticed for your assertiveness and the appreciation that you show for the tough job of running an organization.

IT'S YOUR TURN: RUNNING A MEETING

At some point, either through being noticed by the organization or by your volunteer efforts, you're going to find yourself chairing the meeting. Now is your chance to actually demonstrate some leadership and practice your skills. You should be a chairperson who sends out an agenda 24 hours in advance. In making up the agenda, it is important that you carefully consider each item and allocate on the memo or at the start of the meeting how much time is allocated to each item. *Do not* put so much on the agenda that it cannot all be covered. This will frustrate everyone and make you look inept.

WHAT DO YOU THINK? GETTING OTHERS TO PARTICIPATE

Good leaders are able to motivate people to contribute their ideas and their efforts in order to meet commonly held organizational goals. A good behavior analyst can easily see this as an exercise in shaping (see Chapter 14). Being sure to reinforce people for offering suggestions and making constructive comments will move the agenda along. Leaders not trained in behavior analysis do not realize the value of praise and approval in encouraging more and better ideas and can easily get frustrated. A prompt such as "What's wrong here? Don't any of you have any suggestions for how we can reverse this shortfall?" is unlikely to generate any good solutions. Another strategy leaders who aren't trained in behavior analysis use is to punish good ideas by requiring the person who made the suggestion to follow through. Saying, "Good idea, Jane. Why don't you get in touch with the folks over there at Initech and see if they are willing to cosponsor this event with us?" is more likely to *decrease* her future participation.

YOU CAN'T DO IT ALL: THE IMPORTANCE OF DELEGATING

Learning how to delegate to others is another key skill of good leaders. To be effective, leaders must leverage their position by engaging associates as partners in the enterprise, essentially coaching them to assume their own leadership responsibilities. Learning to make clear the advantages of working on a particular project, along with establishing a history of publicly recognizing people who have taken on additional responsibilities, sets the stage for employees and colleagues to actively volunteer when needed.

LEADERSHIP IN THE WORKPLACE

Because our target readership includes many behavior analysts who are just beginning their career, in this chapter we focused on *leadership skills in meetings* because this will likely be the first place leadership can be practiced. However, behavior analysis as a field is growing so rapidly that many behavior analysts fresh out of graduate school are skipping rungs on the career ladder from therapist to lead therapist to program

> "Behavior analysis as a field is growing so rapidly that many behavior analysts fresh out of graduate school are skipping rungs on the career ladder from therapist to lead therapist to program director while the commencement speech is still ringing in their ears."

director while the commencement speech is still ringing in their ears. Other leadership roles for behavior analysts involve working with families, teachers, teacher aides, paraprofessionals, and nonbehavioral professionals. Leadership skills will also be needed when serving on committees such as human rights committees or behavior program review committees. Behavior analysts who like the idea of working for themselves might elect to own their own company, and they will need visionary leadership to find a niche in the market, find the necessary funding, hire the right people, and direct the consulting firm's activities.

If a behavior analyst is going to be at all successful as a consultant, she must be prepared to take on a leadership role in many situations in a very short time frame. This might involve stepping in to fill a suddenly open slot as team leader or opening a branch office of the firm in a new market in a major city.

INVISIBLE LEADERSHIP

The one trait leaders have that is most difficult to quantify is their ability to come up with the kinds of creative ideas that will catch fire with their colleagues to move the organization in a new direction. All behavioral skills, such as great shaping skills and reinforcement of participation and involvement, will go nowhere if the vision is mundane or flawed. In your preparation for leadership responsibilities, in addition to practicing your committee meeting work, you will find it useful to greatly increase your access to the world of ideas in perhaps unrelated areas. Reading a couple of national newspapers each day, such as the *New York Times* and *USA Today*, in addition to weekly magazines that are "big picture" such as *BusinessWeek*, *Time*, or *Newsweek*, will help provide the perspective you need to think outside the box (for more on this, see Chapter 25, "Aggressive Curiosity"). The ability to do creative big-picture thinking is a characteristic of all leaders, and you can teach yourself to become a more creative thinker (see de Bono, 2008).

SUMMARY

Leaders are intelligent, confident, and enthusiastic. They have mastered critical, practical leadership skills, including running an effective meeting, delegating and giving feedback to others, using excellent social skills, and motivating others to participate in conversations and projects. The development of leadership skills can begin with small steps, such as volunteering to lead a project with a few other people. Over time, by practicing leadership behaviors and continuing to learn about the larger world outside of behavior analysis, leaders develop vision and the ability to engage in creative big-picture thinking.

FOR FURTHER READING

Daniels, A. C., & Daniels, J. E. (2005). *Measure of a leader: An actionable formula for legendary leadership.* Atlanta, GA: Performance Management.

de Bono, E. (2008). *Creativity workout: 62 exercises to unlock your most creative ideas.* Berkeley, CA: Ulysses Press.

Gottfredson, M., & Schaubert, S. (2008). *The breakthrough imperative: How the best managers get outstanding results.* New York: HarperCollins.

Wooden, J., & Jamison, S. (2005). *Wooden on leadership.* New York: McGraw-Hill.

4

Networking

Ron was a well-trained behavior analyst who moved to a big city in a state on the other side of the country. His goal was to work for himself and develop his own caseload of clients. Eventually, he wanted to hire other consultants to work for him.

A year after he relocated, we saw him at a conference and were amazed at how well he was doing. We asked Ron what his secret was. He said, smiling,

> *Two things. First, I'm a good behavior analyst, and second, networking ... networking is what I had to do to be successful on my own. I joined the Chamber of Commerce. I also go every month to our local behavior analysis association meeting, and I go to events where I am likely to meet professionals from the schools, parent groups, and state agencies. I've worked hard at it, but a lot of people know me now.*

As our culture becomes increasingly fast moving, in many fields, behavior analysis included, networking is essential to professional success. If you are currently employed as a behavior ana-

> **"As our culture becomes increasingly fast moving... networking is essential to professional success."**

lyst, there is a good chance you are already networking with colleagues, even though you might not realize it.

For some people, particularly individuals who tend to be social, networking is second nature. If you are a person who likes to meet new people, talk to them, and find out what they do, you're a natural at networking.

Not everyone is skilled at social interactions. You might have noticed that some people in the work setting are shy about introducing themselves. They are not quite sure what they should say or do, and they seem afraid of making a mistake and embarrassing themselves. These tendencies can be troublesome for behavior analysts who are expected to work well with families, other behavior analysts, and professionals from different agencies. The shy behavior analyst should work to overcome shyness in a small, familiar group of people as quickly as possible and then move on to developing a solid set of networking skills.

SO WHAT IS NETWORKING?

Basically, networking is a systematic method of meeting people, finding out about them, and keeping in touch. There may be times when you will be able to provide assistance (in the form of consultations, services, suggestions for resources, etc.) to someone you've networked with, and there will be times when

> "The systematic part of networking involves having an organized plan to find networking opportunities and to seek out people who you think might be able to help you at some point in the future."

people can help you. The systematic part of networking involves having an organized plan to find networking opportunities and to seek out people who you think might be able to help you at some point in the future.

The phone companies understand the concept of networking for families and friends. You've seen the ads about adding to your "circle of friends" and your "favorite contacts." There is a lot to learn about networking, and there are many new skills to acquire if you are going to be successful.

HOW NETWORKING CAN BENEFIT YOU

Let's start with assumptions. We're assuming that you are a recently trained behavior analyst, perhaps starting on your first job in a new city. There is a good chance that you got your job through networking—you knew someone who knew someone who was looking for a behavior analyst, and that person passed the information along to you. You made contact, applied, interviewed, and got the job.

It is estimated that roughly 60% of jobs are landed through networking, so you can see how important this is. Here's another example of having an organized plan for networking. Suppose you are a graduate student taking a course in ethics and professional issues and are interested in acquiring the skills necessary to become a successful consultant. As you approach graduation, you'll want to get to know a lot of people who can spread the word that you are looking for a job and that you are talented, hardworking, and reliable. To be effective in creating your network, you first have to have those qualities, you have to impress some important people, and you need to start building your network of people who will bring up your name if they hear of a position. You'll want to attend conferences to get the word out that you are a well-trained behavior analyst looking for work, and, of course, you'll send letters and e-mail and make phone calls to everyone in your networking circle.

Networking is powered by the magic of *the reference* from someone you know who is willing to vouch for you, and, of course, you must be willing to vouch

> "For networking to have any power, there must be trust and integrity on both sides."

for someone else. For networking to have any power, there must be trust and integrity on both sides. At a recent Association for Behavior Analysis International conference in Chicago, the first author met a former student for breakfast. In a few short years, the student had become a senior consultant at a large consulting firm. As a trusted and well-respected member of the firm, he was now in a position to make suggestions as to whom the firm would hire for new positions. Somewhere around the second cup of coffee, one of the first author's current students walked by. The student was introduced, pleasantries were exchanged, and then he strode off to his table. A brief networking opportunity appeared, and the first author jumped on it: "He is one of my best students, very bright, very reliable, and hardworking, and he shows initiative. I'd recommend him a year from now as a great potential consultant." This kind of input into the network can result in someone getting the job of his dreams; it is effective, however, only when the person recommended performs up to the billing he was given. If you make a mistake and recommend someone who fails to meet expectations, then your credibility goes down—way down.

TWO TYPES OF NETWORKING FOR BEHAVIOR ANALYSTS

Behavior analysts must be prepared to engage in two types of networking. The first will be with nonbehavioral people, and the second will be with behavior analysis professionals.

Networking With Nonbehavioral People

Most often, the nonbehavioral people a behavior analyst will be networking with will be other types of service providers, business owners, or potential clients. Clients may be individuals, agencies, or organizations.

When networking with nonbehavioral people, you'll need to develop a special "elevator speech"

> "You need to be able to explain in 90 seconds or less what behavior analysis is, what you do, and the services that your firm offers."

that is nontechnical and geared toward the average layperson. An elevator speech is one that you could give from start to finish if you had to deliver your message in the time it takes you to go several floors on an elevator. Basically, you need to be able to explain in 90 seconds or less what behavior analysis is, what you do, and the services that your firm offers. Avoid using technical terms, and try to insert at least one memorable, personal anecdote that captures the heart of what you do. Even more important, you need to use your listening and reinforcement skills to find out something about the new contact, discover something you have in common, learn something about the work the person does and the organizations she belongs to, and so on.

Networking With Other Behavior Analysts

The second kind of networking is with other behavior analysts. These are colleagues in your organization or similar organizations in your geographic area. Because they already know what behavior analysis is, your goal is for them to get to know you, and vice versa, so you can create a strong network of people to support you. The kind of support your behavior analyst contacts might provide can range from helping you think through a tough case that has got you stymied, to working out a delicate situation with someone at work, to helping you find a new job.

You might decide that you need to focus on adding some new people to your network. In behavior analysis, we have several ready-made networking opportunities through our local, state, and national associations. For example, in Florida, there are five local chapters of the state association that range in size from 15 to 50 members. These chapters usually meet monthly in a casual environment where a speaker gives a presentation that is followed by a deliberate networking opportunity. At events such as local behavior analysis chapters, leaving immediately after hearing the speaker is a huge mistake. This is a social time where you can practice your networking skills and add new people to your own personal network of friends.

The annual meeting of the Florida Association for Behavior Analysis (held each year in September) draws over 1,000 members from all over Florida and other states. Speakers are invited from all over the country so that behavior analysts interested in net-working can expand their network from those they work with in central Florida to new friends in Michigan, South Carolina, Georgia, Louisiana, California, and Oregon.

Other states such as California, New York, and Texas have similar state associations. At the national level, the Association for Behavior Analysis International meets every year around Memorial Day weekend. The attendance at this conference swells to nearly 5,000 behavior analysts from the United States and many other countries. To describe a large international conference as a *fantastic networking opportunity* is not making an overstatement.

NETWORKING BEHAVIORS

Appearance

Because one of your goals is to make a positive impression on someone, you need to start with your appearance. All of the tips in Chapter 1 on business etiquette apply here. At any networking opportunity, you should be well groomed and dressed appropriately for the occasion (casual for the Behavioral Bash at the Association for Behavior Analysis International conference, business casual for a hospitality suite, formal for a company dinner at an expensive restaurant).

Attitude

If you are new to behavior analysis and one of those people who is not a networking natural, the excitement and anticipation of networking might remind you of when you were a 6-year-old getting ready to attend a birthday party. You don't know what you are going to get into, but you know you

"Put on your big smile, take a few deep breaths, stand up straight, put your shoulders back, and stride into the room, showing confidence and a casual, relaxed attitude."

are going to have a good time. Put on your big smile, take a few deep breaths, stand up straight, put your shoulders back, and stride into the room, showing confidence and a casual, relaxed attitude. Once in the room, stop for a minute to size up the gathering. As a warm-up, you can begin by talking to someone you know. Then make an effort to move around the room and introduce yourself to some new people.

Equipment

You don't need a lot of equipment to be successful at networking, but you will need business cards in a nice card case, a pen, and a small notebook or some 3 × 5 cards.

Most networking events will provide name tags for each person. You might have noticed that people tend to place their name tag on their upper left chest, as though they want to be sure to place their name right over their heart. For networking purposes, we need to start a new trend. *Place your name tag on your upper right chest* so it is in the most logical place to be easily seen when you shake hands with people.

Networking is serious business; you are not in the room simply to smile and meet people and make them feel good. A major goal is to find a few people whom you haven't met before who might become part of your network. These should be people with whom you have something in common and whom you'll be in touch with in the next few days. Get their business cards and possibly some additional information so you can follow up. Asking someone for a business card is considered part of business etiquette. If you haven't done it before, it may seem awkward, but rest assured it is common practice. You don't need to ask everyone for a business card, only those people whom you'd like to contact later.

During a networking opportunity, if you promise a person you will send an article or an e-mail, make sure that you note it on your "to do" list. Following up is the first way for you to prove to the person that you are reliable and dependable. Many people are so poor at this that it is likely that the person will be somewhat surprised that you actually remembered what you promised.

Specifics of Behavioral Networking

As you approach someone, put on your big friendly smile, throw back your shoulders, put out your hand for a handshake, and introduce yourself. Then ask a noncontroversial and open-ended question that will get the conversation started: "This is a great event; how did you hear about it?" "What business are you in?" "This is a wonderful conference. Did you hear any good papers today?" "I love your tie. I've been looking for something like that for my husband." The old rule about staying away from politics and religion is a good idea and can prevent you from offending someone. Once you get the person you'd like to meet to start talking, you can be reinforcing and demonstrate your great listening skills. This should be sincere, natural, and not contrived. If you're a behavior analyst, we hope that you like people and will truly come to enjoy hearing a new friend open up and talk about hobbies, a recent trip, or a particular area of interest in behavior analysis.

Being a good listener is important, but you don't want the interaction to end with your being such a great listener that you never said a word. When the time is right, look for your opening to say something about what you do. Remember that you're looking for some common ground, something that might link this person to you or to someone else you know. You might not be looking for a job, but a friend or acquaintance might be, so you could start your networking by helping someone else.

Two rules will help guide your activities for any event: Be curious, and be a connector.

Be Curious   Don't be afraid to ask questions about what people do and their hobbies, travel, children, or dogs. If they look like they are changing the subject or choosing to not elaborate on a particular topic, you need to *pick up on the subtle cues and respond appropriately.* For example, most people enjoy talking about their families. There will be situations, however, where there is dysfunction in a family, and your new friend might not wish to tell you that her children are losers, her mother is a punishing and sarcastic nag, and her husband just ran off with his secretary.

Be a Connector   As a good networker, you should try to put people together whenever you can. If you are successful, both parties will be thankful and will at some point return the favor.

During your conversation, if you realize the person you are talking to should meet someone else in the room, by all means offer to make the introduction. Say, "I have someone I'd like you to meet," and take the person over to your friend. Give a good introduction to each: "Sarah, this is Charlotte. She's the new HR director at Bright Kids." Your goal is to make a friendly impression, find a connection for yourself or someone else, get a business card, and move on. You should not dominate the time of the person you just met, because you have many more people to meet and so does everyone else. To end the conversation, be as polite as you were in the beginning, and remember that last impressions count too. To end an interaction, you could say, "I really enjoyed meeting you. Have a great time at the conference." Ideally, you'll be able to include someone else in the conversation so you don't have to leave a person totally alone.

Networking Follow-up

This aspect of networking is not discussed very often, but this is where the payoff is for the whole networking enterprise. After the networking event (e.g., social hour, informal time to chat after a meeting), such as later that evening or early the next morning, review the business cards you accumulated and the notes you took. As you flip through each one, recall the conversation and decide if there was a potentially valuable connection for either yourself or someone else. If you promised to send a person a link to a Web site or an article to read, definitely follow through with your promise. In some cases, you'll want to follow up by phone;

otherwise an e-mail will do. For a connection that looks promising, an e-mail suggesting a lunch meeting would be appropriate. Be sure to remind the person where you met and what you think the connection is: "We met at the Bay Area Association for Behavior Analysis on Tuesday night, and I am hoping we can continue our discussion of …"

Virtual Networking

Perhaps the most intriguing form of networking is the virtual kind. This is where you meet someone online, through a chat room or a blog, and then follow up via instant messaging, text messaging, or iChat (Apple Inc.'s video online system).* If you use Google, you can customize your start page so it searches for key words on blogs in which you might be interested. If you get a hit, you can follow up with the person and strike up a conversation that might lead somewhere.

Intentional Networking the Modern Way

Most of the networking we have been describing has been of the chance variety, as in the situation in which you go to an event where you hope to run into someone who might be a great connection. The networking is iffy, and you might go to four or five meetings or conferences before you find a good match with someone who has common interests. Another option is Internet-based, and this method is much more likely to produce some contacts. These are Facebook-style professional (as opposed to social) Internet sites where people sign up specifically to network with like-minded souls. LinkedIn is one example of such a site (www.linkedin.com). Facebook is the original networking site, and it offers business and professional groups as well. You can create a network for your organization by going to www.ning.com, where you can customize a home page to suit your needs and invite your colleagues or anyone with similar interests to join. Again, we are talking about professional interests here, such as interacting with other behavior analysts who work with preschool

---

* Go to www.apple.com/macosx/features/ichat.html for more information.

children with autism. There are plenty of Internet sites for the purpose of helping you find the love of your life; this section is about networking related to your career.

Payoff

The goal of networking is to find people with common interests who can support each other. If you can do this, you will be sought out, and your connections will make a difference in people's lives.

> "The goal of networking is to find people with common interests who can support each other."

Here is an example. The first author read an article in the local paper about some research on human behavior being conducted at the airport. He used Google and other sites on the Internet to track down the company and person named in the article. He then called the principal investigator (Internet networking). An appointment was set up for a face-to-face meeting, and a personal connection was made. This resulted in an offer to assist in the research effort. The first author had a graduate student who needed a dissertation topic, and the student became involved in the project. Further meetings were held to seal the deal. Many months of writing proposals and gaining Institutional Review Board (IRB) approval ensued, followed by many more months and hundreds of grueling hours of direct observations, including a late night shift for data collection that ended at 2:00 a.m. As the research began to yield results, networking with others in the field was initiated. This resulted in contact with a major consulting firm in Washington, DC, that conducted federally funded, high-security research on human behavior. The principal of that firm turned out to be a long-lost colleague of the first author (delayed networking outcome). The long-lost colleague was invited to give an address at the university, where a direct connection with the graduate student took place (excellent face-to-face networking). One year later, as the student was defending his dissertation, the consulting firm had a job offer for the new PhD. Networking of all kinds, from skimming the paper to sending e-mail and making phone calls to making in-person visits, paid off in a big way to create an amazing research opportunity and a chance-of-a-lifetime job.

Networking Ethics

Unfortunately, some people in this world are users. These are the people who will engage in networking in professional settings so they can try to gain access to their own personal reinforcers, such as free meals, wine, or tickets to sports or cultural events. This is not the purpose of professional networking, and it should be avoided.

SUMMARY

We can use networking to meet new colleagues, advance a career, gain resources and services for clients, introduce nonbehavioral people to our field (remember to have an elevator speech ready), and promote an organization. Networking is the skill that helps us become a part of the behavioral community.

FOR FURTHER READING

Darling, D. C. (2003). *The networking survival guide: Get the success you want by tapping into the people you know.* New York: McGraw-Hill.

5

Public Relations

Robotic behavior, lack of emotion and inability to use trained skills outside school are some of the shortcomings critics attribute to ABA [applied behavior analysis]. A boy who has learned to play Nintendo games at Alpine, for instance, reverts to simply switching the game on and off when at home. ...

CTC [Celebrate the Children] emphasizes the expression of emotion and spontaneous thinking. Rather than work on a highly specific skill, DIR [developmental, individual-difference, relationship based] activities tend to include complex social interactions that build many skills at once.

Claudia Wallis

Public relations—the means by which we manage the flow of positive information—and behavior analysis are practically strangers. As a field, we have done an absolutely atrocious job of reaching out to the public with our story. The above quotes from a 2006 article in *Time* show the devastating results. Without question, *Time* is one of the most well-known American news magazines. In addition to its readership of more than 4 million readers per week in the United States, *Time* reaches a global audience.

Even though a behavioral program may be excellent, when a news publication with the stature of *Time* reports that the impression behavior analysts give our target audience is that we are cold, we lack emotion, and our behavior analytic methods produce robots, something is seriously wrong.

Letting the community know what we do, of course, is vitally important to the advancement of our profession. We need broad public acceptance if we are to truly have an impact on the cul-

> "We need broad public acceptance if we are to truly have an impact on the culture."

ture. The next generation of behavior analysts must be prepared to accept the challenge of increasing our media exposure. Every day, stories appear in newspapers and magazines, on television, and on the Internet about other approaches to human behavior that are not in the least data based, and we often get slighted or misrepresented by our competition. This is certainly to our detriment. Public perception of behavior analysis is critically important, because most people learn about us through the media rather than through scholarly journals, textbooks, or college courses. A single article, such as the one in the May 2006 issue of *Time*, can reach over 4 million readers, resulting in 4 million people who might now have a total misconception of our field and what we can offer. It's time to fight back, but we can do so only if we understand our public relations goal and if we have a plan to reach the citizenry.

Most graduate programs do not prepare students for media exposure or explain how the process of contemporary public relations works. Historically, reporters and journalists kept their ear to the ground and looked for interesting developments in their communities, searched out stories, tracked down sources, and conducted interviews, and when they found something they thought their readers would be interested in or, more important, found an intriguing bit of information that might shock or enlighten the populace, they would write about it and bring it to the public's attention. Those days are over. The real world of media coverage is much different. Reporters tend to stay in their offices, inundated by press releases and overwhelmed by phone calls and e-mail that provide leads on stories that have been skillfully prepackaged by publicists. When reporters find something that suits their fancy, they might conduct phone interviews or engage the principals in an e-mail exchange. On rare occasions, they might actually visit a site, conduct face-to-face interviews, and really dig into a story.

To crack this system and get the word out about what we do, the successful behavior analyst needs to have a plan of action. Two professional, academic behavior analysts, Drs. Sharon and Ken Reeve, from the ABA faculty at Caldwell College in New Jersey,* have been very successful in getting accurate media coverage for their program and have provided some excellent tips for behavior analysts.† The model for behavior analysis public relations they developed is so outstanding that, with their permission, we included the majority of their suggestions in the following list. They start by using a free Internet tool that is available to everyone, Google.

- The first thing you must do is set up a Google Alert (www.google.com/alerts). This service provided by Google allows you to tell it what key words you want to hear about from the news, blogs, or Web sites. It will then e-mail you links to any Web-based mentions of those key words. Many of our strategies described later are aided by our receiving Google Alerts.
- Do a search for yourself on Google every day to see if you or your program is mentioned. Sometimes we find someone is discussing in a blog something we wrote or said. When we find it, we can then comment on what is being said right in that blog. On one occasion, some parents were debating the merits of getting training in ABA, and Caldwell College's ABA program came up. Unfortunately, it was mischaracterized (although not in a mean way).

---

* Drs. Sharon and Ken Reeve can be reached at Psychology Department, Applied Behavior Analysis Graduate Programs, Caldwell College, 120 Bloomfield Avenue, Caldwell, NJ 07006; phone: (973) 618-3315, office 6-330; http://faculty.caldwell.edu/sreeve.
† See "A Master's in Self-Help," *New York Times,* April 20, 2008, p. M213.

We just wrote in our own comments about what was being said. This directly led to two parents deciding to enroll in our program.

- Get involved in local ABA chapters and local autism groups. You will be amazed at the number of important connections you can make while serving the professional and parent communities. Often, you will serve on committees that might affect public policy and that might gain the attention of news sources.

- If you hear about any autism summits (or find out through Google Alerts), try to get yourself invited so that the ABA side of things is represented. Oftentimes no one is at these meetings to promote evidence-based interventions. People will want to listen to what you have to say. Of course, if you have other special interest areas such as behavioral safety, you'll adapt this approach to keep track of your interest area.

- Invite yourself to any press conferences government officials will be giving. Introduce yourself to the important attendees. You can often get invitations by giving your credentials to the official's assistant. We currently have relationships with a number of local officials as a result of this tactic.

- If you run a quality program, such as a school or autism program, invite public officials to visit.

- Keep an eye on the Web pages of local politicians to see if they are supporting any autism legislation. If they are, try to get in touch with them and offer services as an expert. You might get a chance to change public policy. It also helps to check the Web pages of quality advocacy organizations, such as COSAC, which keeps an ongoing description of autism and ABA-related legislation on its Web page.*

---

* The New Jersey Center for Outreach and Services for the Autism Community (www.njcosac.org).

- Present talks to as many parent and support groups as you can. Parents are often starved to hear from a professional, and many of them have contacts in radio, TV, and newspapers. If you prove yourself worthy, you might get recommended to tell your story. Give out your bio when you do these presentations.
- Offer to give a talk at your local library about autism treatment and ABA. This request is especially likely to get a positive answer during April, which is Autism Awareness Month.
- Write a resource pamphlet for your local library. We did this and then were asked to work on a grant with the library on how persons with autism could better access library resources. This led to our being asked to give talks at that library and others in the surrounding towns. These talks were highlighted in our local newspaper. People then were calling the college to inquire about our ABA program.
- Within your city or town, research the archives of the newspapers to see if there are any stories on ABA or autism. If so, contact the reporters who wrote these stories, compliment them on past columns, then pitch some new stories to them (that's what we did to get our story in the *New York Times*).
- Set up for a popular audience your own Web page or blog that has links to resources and descriptions about what ABA is. Also include a current bio of yourself for any interested news media people to see (and to use to refer to you in any stories they do on you). If you belong to a university, it might allow you to do this on your faculty Web page.
- Write articles for popular autism magazines such as *Autism Spectrum Quarterly* or *Autism Asperger's Digest Magazine.*
- Give a workshop at a local university. Most universities will rent space, and you get to keep all proceeds after you pay

for the space. Some universities may even let you use the space rent free if you convince them the workshop features a public service topic. In this case you would not charge the participants a fee.

- Suggest a course on ABA and autism to your local college. Offer to design the curriculum for the class. Pitch this to the psychology or special education department chair. Then teach the class, and teach it well. Ask interested students (50 is a good number) to write the college president and suggest the college offer a certificate program in ABA. Offer to design the program for the college with help from colleagues who already have a program approved by the Behavior Analyst Certification Board (BACB). If the school approves, ask about the possibility of getting a faculty position.

- If you work at a college or university, establish a relationship with the public relations person or Institutional Advancement office to discuss how autism treatment issues can be used to market the college. Organize student autism walks. Write stories on ABA or autism for the college newspaper. Work with the Psychology Club to have it do service projects for local autism schools.

- Organize a grassroots mass mailing (with as many colleagues, friends, and family members that you can) to a local newspaper asking it to cover a particular story, conference, or event.

- Go to Amazon.com and make lists for recommended reading. One reporter we spoke to had read Dr. Ken Reeve's list that he posted on recommended ABA and autism sources.

- Write a bio for popular consumption with bulleted accomplishments and statements about what you can do: "I can talk about finding effective autism treatments," "I can separate fad treatments from real ones," or "I can provide a description of applied behavior analysis treatment."

- Media people have told us they prefer to have people with media experience on as interviewees, so you should include in your bio all your activities, with specifics such as "Interviewed by WFFM 1020 on August 10, 2008, regarding vaccine controversy."
- Once you get something published in a popular medium, you should use copies of it as a "takeaway piece" to give to anyone who is interested in what you have to say but who cannot spend much time with you (reporters on the run, overworked legislative personnel). We did an article on autism and ABA for an alumni magazine, and it is our standard takeaway piece, because it mentions our work in developing Caldwell's ABA program.
- In preparation for working with reporters, learn how to speak in sound bites. Come up with user-friendly, 20-second descriptions of what autism is, what ABA is, what evidence-based treatment means, why ABA is medically necessary, and why it should be funded through health insurance. For example, "In applied behavior analysis, we break skills down into easily learned parts, give children a lot of guidance and positive feedback while they are learning, and make frequent observations to make sure what we do is effective." Videotape yourself, and ask colleagues to critique you.
- If a reporter misquotes you or puts the wrong spin on your story, do not let it stop you from continuing to fight the good fight. We have now appeared in print a number of times and have never been completely happy with the results. It seems, however, that the more we effectively translate ABA to a popular audience, the less likely it is that we will be mischaracterized.
- When you establish a relationship with reporters, offer to give them advice on other contacts or resources in the field. Someone often has a relative with autism.

- And finally, *never* blow off an appointment with a reporter. This will result in your name being put on the "Do Not Contact" list.

Dr. Bobby Newman is another behavior analyst who has had a great deal of experience in representing ABA to the general public.[*] He generously offered some suggestions for presentations to the general public.

- Know your audience, and act and dress accordingly. The same suit that will get you respect from one audience may be off-putting to another. Wearing jeans and talking about injuries you've received over the years during clinical work and using everyday vocabulary is quite appropriate for an after-work presentation for some groups of paraprofessionals, for example.
- Be prepared to discuss the scientific literature in a professional manner when this is appropriate, for example, when talking to parents who are well educated about ABA.
- Use stories and case studies. They are more convincing to nonprofessionals than graphs and data. When presenting alongside parents, emphasize that you bring the literature and methodology to the table, and they bring the specifics about their child. You could never hope to know what they do, and this is a partnership.
- Be real, and don't be afraid to show emotion when celebrating successes or lamenting previous unspeakable conditions. If you're giving an example based on things your late father told you, don't be afraid to show some emotion about how you love your dad.
- Don't be afraid to share personal info about your life. Understand that the audience members are going to draw a picture of you one way or another in their head. They can't project images onto a screen that isn't blank.

---

[*] Dr. Bobby Newman can be reached at www.room2grow.org.

- Take the subject matter seriously, not yourself. Feel free to make fun of yourself and point out silly things you have done in your nonclinical life.
- Emphasize common goals, then talk about ABA as a way to achieve them. Always keep an eye on common interests and goals, not personalities.
- Make frequent reference to the BACB code of ethics. People need to know why you are doing what you are doing and that there is an ethical authority for your practice.
- Don't be afraid to inject humor into examples; feel free to make fun of your own reinforcers, for example.
- Be aware of common objections to ABA, and be prepared to effectively argue against them, emphasizing how ABA is meant to build individual autonomy and choice.
- Speak warmly of the people who trained you and of other professionals in the field whom you respect. Speak especially warmly of consumers and clients whom you truly respect.
- Try to represent the field in a way in which everyone can be proud, including being a responsible citizen of the greater world. Every February, for example, I put together "Team ABA," which does a polar bear swim and raises money for the Make-A-Wish Foundation. We need more such efforts to show that we are caring members of the greater world.

HELP FROM YOUR NATIONAL ASSOCIATION

Although you can be quite successful with many of your own public relations efforts, there will be times when the issue is just too big for an individual behavior analyst. Remember that your national professional organizations have a responsibility to help with public relations for the field of behavior analysis. The Association of Professional Behavior Analysts and the Association for Behavior Analysis International are in a position to send press releases and media alerts when public relations efforts are needed on a large scale. If you are a member of these organizations, you can request their assistance in public relations.

SUMMARY

As behavior analysts, we have quite a challenge ahead. Our competition, the individuals and organizations pushing snake oil, questionable practices, and outright hoaxes on the American public, is getting a lot of play in the media. For the most part, the advantages these individuals and organizations have with the press are that they are new and have no qualms about offering amazing and fantastic results. To clear up this haze of misinformation, we need to work hard to reach out to the media in all the ways that Sharon and Ken Reeve and Bobby Newman suggested. If you had a child with a disability, think about the services you would search for and ultimately choose. Although we need to remain committed to our evidence-based, scientific approach, we need to emphasize our underlying core goals of an improved quality of life, more choices and freedom from restrictive consequences, and, most of all, independence and autonomy for our clients.

FOR FURTHER READING

Hall, P. (2007). *The new PR: An insider's guide to changing the face of public relations*. Potomac, MD: Larstan.

Laermer, R. (2003). *Full frontal PR: Building buzz about your business, your product, or you*. Princeton, NJ: Bloomberg Press.

6

Total Competence in Applied Behavior Analysis and in Your Specialty

Blood was streaming out of Tim's ear and down the side of his cheek. It was then I realized I was in over my head.

Anonymous, a new MA-level Board Certified Behavior Analyst®

Behavior analysis offers us a set of basic principles that explain a wide range of behavior. Our field is sufficiently robust in that it offers treatments for the complete spectrum of clients, from infants to seniors, from individuals with severe and profound disabilities to world-class athletes and Fortune 500 CEOs. The *Journal of Applied Behavior Analysis* (JABA) has provided us over 40 years of well-controlled studies that point to a precise methodology for measurement, a complete system for functional analysis, and a replicable and reliable treatment protocol that fulfills B. F. Skinner's dream of a validated science of behavior that can be put into service for the benefit of mankind.

This strong, scientific foundation obviously puts us in an ideal position to make a significant contribution to society. The downside is that in some circumstances, we might raise the expectations of our consumers and not be able to meet them—not because we don't have the technology but because our personnel are not sufficiently versed in the details for each population.

In short, problems arise when behavior analysts are not totally competent in an area for which they are asked to provide treatment. A behavior analyst who specialized in autism treatment in graduate school may not know exactly how to work with patients with Alzheimer's disease, and someone who got his practicum training in an elementary school classroom may not know how to treat a teenager with developmental disabilities and serious self-injurious behavior. Our Guidelines for Responsible Conduct make it clear that behavior analysts should not practice outside their area of expertise, but this doesn't prevent administrators, supervisors, parents, or others from asking for help with such cases.* The caring and compassionate behavior analyst might have a hard time turning away such requests, especially if there do not appear to be any other qualified individuals around.

TOTAL COMPETENCE IN APPLIED BEHAVIOR ANALYSIS

To be totally competent in applied behavior analysis means that you have gone beyond the minimal requirements to pass the Behavior Analyst Certification Board (BACB) exam. Way beyond. It is widely acknowledged that the exam is a test of *minimal* competence to practice in the field. There is an understanding that board-certified *assistants* should be supervised by Board Certified Behavior Analysts®. Passing the BACB exam, however, does not mean that you are qualified to take on *any* case that comes along. The expectation of leaders in the field is that you will continue to read on your own and participate in continuing education as long as you are a member of this proud new profession.

Expanding your knowledge of behavior analysis means keeping up with new books that are coming out each year, going to the

---

* See J. S. Bailey and M. R. Burch, *Ethics for behavior analysts: A practical guide to the behavior analyst certification board guidelines for responsible conduct*, Mahwah, NJ: Lawrence Erlbaum Associates, 2005, pp. 227–252.

annual Association for Behavior Analysis International (ABAI) conference, and attending sessions both in your specialty and in broader scope areas. The ABAI has initiated the B. F. Skinner Lecture Series in which scholars and researchers indirectly related to behavior analysis are invited to present their latest data and their innovative theories related to behavior. For example, at the 2008 conference, you could attend sessions titled "Empirical Measures of Social Information Processing," "Neurobiology of Cocaine Self-Administration," or "Not by Genes Alone: How Culture Transformed Human Evolution." Each of these sessions was a mind-expanding, enriching experience where behavior analysts in attendance were challenged to formulate a world-view of behavior analysis that is intriguing, complex, and exciting. To think that our perspective on behavior is taken seriously by behavioral scientists of all types is gratifying, and to begin to understand how everything fits together is challenging, to say the least. Attending conferences and attending sessions will advance your competence to a new level.

Another way to stay competent in behavior analysis is to engage in regular dialogues with colleagues on topics of the day. You can take your cue from articles in *The Behavior Analyst, The Analysis of Verbal Behavior, JABA*, or the new journal *Behavior Analysis in Practice* and gather with colleagues to discuss and debate topics such as "the nature of clinical depression" or "culturally sensitive functional analytic psychotherapy." It is through such discussions that you can remain sharp and current on the theoretical issues that are fundamental to our field. Keeping up with applied research involves revisiting your research methods class notes and taking a shot at digesting topics such as the "utility of extinction-induced response variability for the selection of mands." Experts in the field read the professional journals regularly, ponder the articles, and think of ways to apply the theories and data to their everyday practices. If you want to someday be counted among the experts, you'll work to develop this very valuable habit.

TOTAL COMPETENCE IN YOUR SPECIALTY

Within applied behavior analysis, there are a number of specialty areas. Starting with autism and moving on to the work that is done in zoos, we can accurately say that behavior analysis specialties range from A to Z. Specialty areas within behavior analysis include but are not limited to areas as diverse as animal behavior, autism, behavioral gerontology, behavioral medicine, behavioral safety, clinical and developmental behavior analysis, developmental disabilities, direct instruction, evidence-based practice, the experimental analysis of behavior, health and sports issues, organizational behavior management, parent training, performance management, sex therapy, and verbal behavior.

Total competence within a specialty means that you have first defined your area of expertise as it relates to the course work you've taken and the practicum experience you've had. Total competence also requires that you complete the necessary continuing education hours and read additional

> "Total competence also requires that you complete the necessary continuing education hours and read additional materials that will help you keep up with your specialty."

materials that will help you keep up with your specialty. If you want to be competent within a specialty area, you'll probably need to subscribe to two or three specialty journals in addition to your ongoing general reading. Finally, to develop total competence, you will need years of experience working in the field under a range of circumstances from well-controlled clinic settings to community placements.

Within your specialty area, as you gain years of experience, you will be able to determine which cases you can safely take and those that you need to refer to another professional. The following list can be used to determine if you are totally competent in your particular area of expertise:

- I have read the most recent studies in the best journals on this topic.
- I have on my bookshelf the books that are considered the landmark works on this topic, and I have read them thoroughly.
- I have attended workshops on this specific topic in the past year.
- I was personally trained by a Board Certified Behavior Analyst® who is well qualified in this specialty.
- I can identify the experts on this topic. I have met them, and I can reach them by phone if necessary.
- In my consulting practice, colleagues regularly consult me on cases involving my specialty.
- I have given presentations on this topic at state or national meetings of behavior analysts.

MONITORING YOUR OWN PRACTICE

In addition to keeping up with your reading and going to conferences, the next most important way that you can remain totally competent is to carefully monitor your own practice. This involves conducting an ethics check prior to accepting each case (see Chapter 7) and at each step throughout the treatment process. You must collect data all the way from baseline to follow-up so you can determine if your behavior program is effective. This is a unique requirement of our field (Guidelines 2.09, 4.04) that clearly separates us from the rest of the human services, and we want to make sure that we fulfill it with integrity. Seeking out colleagues who will review your data is another important aspect of competence. This falls into the category of peer review. It is one thing to look at your own data and pronounce it "solid," and it is another thing to show it to an independent professional for an opinion. If you have an organized peer review committee in your area, you can use its members' oversight and feedback to maintain your skills. Peer review by other competent behavior analysts will ensure that you stay on the cutting edge of our field. You may even be able to raise the bar for your colleagues by educating them about some new techniques you learned at a conference or read about in the latest issues of *JABA* or *Behavior Analysis in Practice.*

KNOWING WHEN TO DECLINE A CASE

In the opening quote, "Anonymous" realized he was in over his head when he observed blood dripping from his new client's ear. He had taken a case involving a client with self-injurious behavior (SIB) without realizing how serious it was. Dangerous situations such as this can happen when eager, young behavior analysts try to prove themselves and accept a case without examining the intake information closely and without objectively reviewing their own competencies.

A one-semester practicum in a group home for ambulatory clients with severe disabilities is not enough preparation to deal with profoundly involved individuals who are in a residential setting. SIB in the former setting might involve clients who occasionally bite themselves or threaten to hit their heads. But when a profoundly involved client is in a residential setting, it usually suggests that the behavior problems are more pronounced and that the client may have a history of extreme behaviors. The behavior analyst (in the quote at the beginning of this chapter) found out only after starting this case that Tim (the client) banged his head with his fist and that at one point in the recent past had put his head through a window. He was taken to the emergency room for nearly a dozen stitches.

Our new behavior analyst put himself in a liability situation where, under some circumstances, his professional career could have come to a screeching halt. Having *total competence in your specialty* means understanding the complexities of behavior for your area of expertise. This includes knowing what to look for in a client folder, including the medical section, and understanding enough of the medical terminology to determine the severity of the condition. Grasping the technical difference between *lacerations, abrasions,* and *contusions,* for example, and having the professional background to inquire about the client's most recent behavioral incidents, as well as the conditions under which the injuries occurred, indicate that you are fully aware of what you are getting into when you take a case. This level of competence is needed not only for self-preservation but also to ensure the best interest of your client. If a client is injured on your watch, you must take some responsibility. If it turns out that you really were not qualified to handle the case, this could come back to haunt you.

> "If a client is injured on your watch, you must take some responsibility. If it turns out that you really were not qualified to handle the case, this could come back to haunt you."

DEVELOPING COMPETENCE IN A NEW AREA

In a case related to competence in performance management (PM), a brand-new master's-level behavior analyst with some basic experience in PM had an opportunity to interview for a job involving behavioral safety in open-pit mining and the steel industry. The job was in another country, and the graduate sought advice from her major professor. "I don't know anything about mining or steel production, but I've always wanted to travel, and this job is perfect for me," she said excitedly. After reviewing the job description, the major professor asked, "What they are looking for is someone who learns quickly and who can communicate with people who wear steel-toed boots and hard hats. Can you do that?" "Absolutely," was the response. "OK, then accept the interview, and emphasize that you are a quick study and that you love meeting people. Talk about the experience you've had in a variety of business situations. And don't forget to mention that you prepared for 6 months to compete in a fund-raising marathon, finished in the top 100 runners, and raised over $3,000 for charity. Finally, don't forget to ask questions about how you will be trained and the nature of your supervision."

The new graduate landed her dream job. Even though she was not competent in the area of open-pit mining, she felt comfortable with accepting the job because her consulting firm made it clear before she started that she would receive extensive training and supervision. In 6 months, she was fully responsible for two clients and was ready to train another new consultant with the firm. She reported, "They gave me two 4-inch binders of information and told me I had to be fluent on the material in a week. I was able to demonstrate that I'm a quick study, and my graduate school training really paid off in this situation."

SUMMARY

Becoming competent in applied behavior analysis and your specialty area is critically important to maintain the standards of excellence for our field. Reading the literature, attending conferences, interacting with behavioral colleagues, systematically developing all of the skills relevant for your specialty area, and passing the BACB certification exam will ensure that you remain proficient in your professional practice.

FOR FURTHER READING

Behavior Analysis in Practice. Kalamazoo, MI: Association for Behavior Analysis International.

Behavior Analyst Certification Board. (2004). Guidelines for Responsible Conduct. Tallahassee, FL: BACB.

Behavioral Interventions. New York: John Wiley & Sons.

Journal of Applied Behavior Analysis. Bloomington, IN: Society for the Experimental Analysis of Behavior.

Journal of Organizational Behavior Management. Philadelphia, PA: Taylor & Francis Group.

Skinner, B. F. (1953). *Science and human behavior.* New York: Macmillan.

7
Ethics in Daily Life

---

*After a few weeks on the job, it became clear to me the data that
staff were turning in were made up. The supervisor denied there
was a problem, and the administrator was no help at all.*

**Anonymous, an MA-level Board Certified
Behavior Analyst® with 1 year of experience**

Every day, behavior analysts, in the course of performing
their work, confront ethical problems. In the course of
addressing ethical issues, behavior analysts must make decisions that can ultimately produce serious consequences for
their clients, third parties, and themselves.

You may have experienced situations such as a curious teacher
asking you about confidential information that pertains to a
child's family, a colleague routinely billing for more hours than
he works, or a client's parent giving you an expensive bottle of
wine as a thank-you for the services you are providing, or you
may have discovered that data (taken by others) used to make
important decisions about a client's behavior plan are not valid.
How you respond to these and countless other ethical situations
that come up every week can put you in a situation where your
most important concern to "do no harm" is challenged.

In *Ethics for Behavior Analysts* (Bailey & Burch, 2005), we provided detailed explanations of the Guidelines for Responsible

Conduct for behavior analysts. If interpreted properly and followed faithfully, these guidelines should protect you and your clients. As we conducted daylong ethics workshops throughout the United States, we've noticed some frequently reported ethical issues. If not handled properly, these issues, listed next, can result in serious complications for the behavior analyst.

DAILY ETHICAL CHALLENGES

Integrity

As discussed in Chapter 5, one of the most important challenges to the daily conduct of the professional behavioral consultant is the integrity of the services provided (Guideline 1.05). If we are to instill confidence in our consumers, it is essential that they come to see us as truthful, reliable sources of information about behavior and how it can

> "Ethical behavior analysts do not promise what they cannot deliver, shade the truth, make outlandish claims, or waver from a commitment."

be changed. Ethical behavior analysts do not promise what they cannot deliver, shade the truth, make outlandish claims, or waver from a commitment. They certainly *do* obey the law. Promises matter, and the extent to which a behavior analyst can consistently deliver exactly what was promised on time, and with no excuses, is the extent to which the behavior analyst will be seen as ethical. A person with integrity is one who cannot be persuaded to take a different position just because it is popular, easy, or even rewarding in some immediate way.

One current dilemma presents an excellent example of a challenge to the integrity of behavior analysis services, and that is the plethora of "alternative" services offered to parents of autistic children. Behavior analysts tell us they are asked almost daily by parents or teachers about hyperbaric oxygen treatment, gluten-free and casein-free diets, sensory integration, auditory integration, and the like. Parents want your opinion and often want you to tell them it is OK to engage the services of someone offering these nonevidence-based treatments. Our code of ethics says, however, that you must inform the client about our firm commitment to data-based approaches. *Integrity means upholding high moral principles*, and in this regard we have three: (a) a responsibility to all parties affected by our services; (b) a commitment to evidence-based treatments; and, above all else, (c) do no harm.

Competence

As described in Chapter 6, the primary concern of behavior analysts is that they not practice outside of their area of competence (Guidelines 1.03, 3.04). Exactly how one determines the "boundaries of competencies" is not spelled out in detail in the guidelines, but it is assumed that a conservative approach is best. If these guidelines are followed religiously, careful behavior analysts will risk no harm to clients or their own professional reputation.

Confidentiality

In the course of daily interactions with clients, the behavior analyst comes into contact with a great deal of information of a private nature about the clients. Oftentimes, through having conversations, reviewing records, or making direct observations in the home, the behavior analyst can determine the intimate details of a person's life. If shared with others, this knowledge could destroy important relationships or damage a person's career or position in the community. For this and other reasons, behavior analysts are admonished to "take reasonable precautions to respect the confidentiality" of the clients with whom they work.

Dual Relationships

The very nature of our work with clients puts us right in the middle of their life. We help people adapt to complex environments, teach them more appropriate responses, arrange contingencies so they get more out of life, and improve their relationships with others.

For this, the clients and their families and surrogates are often grateful and want to show appreciation in tangible ways. This may include offering gifts or invitations to participate in family events such as birthday parties. Becoming a friend of the client or the client's family is often the start of a slippery slope that can ultimately undermine the behavior analyst's professional judgment. Guideline 1.07 advises against entering into such dual relationships because of the possibility that it will impair the behavior analyst's judgment or interfere with "the behavior analyst's ability to effectively perform his or her functions" (Bailey & Burch, 2005, pp. 229–230).

Functional Assessment

One scenario in our ethics seminars that often catches people unaware (Bailey & Burch, 2005, p. 269) begins with the following: "Kevin continues to bang his head when attempting to seek attention from his parents and teachers." The scenario then describes a series of other treatments that have been tried, including "deep pressure" from an occupational therapist and sign language recommended by the speech therapist. Behavior analysts are asked, "How long should interventions continue to be in place before medication or shock therapy is considered?"

In a sense, this is a trick question, because a careful reading indicates we don't know the function of the behavior. "When attempting to seek attention" is just a description, therefore, it is inappropriate to continue *any* of the treatments.

In most cases a referral comes with a suggestion for an intervention: "My 2-year-old whines and cries and ruins our dinner when we are out with friends. What can I do to make him stop?" There seems to be hope in this question that the behavior analyst can provide some magical, easy consequence to stop the whining and crying even though the *cause* of the behavior is not known. Saying, "Try this ..." would be highly inappropriate and unethical. The child could be crying for attention, the dinners could be adult oriented and totally inappropriate for the developmental level of a 2-year-old, or the child may be trying to say, "I am exhausted from being dragged to restaurants for long dinners."

Right to Effective, Least Restrictive Treatment

The data-based treatment approach used in behavior analysis puts us in a unique position in relation to other human services. On an almost daily basis, the behavior analyst consultant will be confronted with a decision on the most appropriate treatment for a client where the alterna-

> "On an almost daily basis, the behavior analyst consultant will be confronted with a decision on the most appropriate treatment for a client where the alternatives involve fad treatments, warmed-over placebo effects, or outright frauds being pushed as though wishful thinking and hope were all that mattered."

tives involve fad treatments, warmed-over placebo effects, or outright frauds being pushed as though wishful thinking and hope were all that mattered. In our field, we take a hard-line stance on such fly-by-night nostrums, potions, and elixirs, and our guidelines indicate that we "are responsible for review and appraisal of all alternative treatments, including those provided by other disciplines" (Bailey & Burch, 2005, p. 233). As behavior analysts, we have to ensure our treatments are grounded in peer-reviewed applied research (Bailey & Burch, 2005), and they have to be constantly evaluated while the treatment is in place (Guideline 4.04). Our guidelines further specify that we avoid harmful reinforcers (Guideline 4.03), recommend reinforcement instead of punishment (Guideline 4.02), and eliminate those conditions that might hamper the proper implementation of behavior programs (Guideline 3.02). This latter guideline is an essential part of the ethical conduct of professionals in our field. Because we rely on mediators to carry out the programs we write, we need to make sure that they are qualified to do so (Guideline 5.08). We also need to train them competently and monitor their performance (Guidelines 5.09, 5.10).

ONGOING DATA COLLECTION

Our origins in the experimental analysis of behavior have given us a theory of behavior that says most behavior is learned through consequences in the environment. From our experimental roots, we have also developed a methodology of measurement and intervention. One distinctive aspect of this methodology is the requirement for ongoing data collection. Once a stable baseline has been collected and the intervention begun, the data collection does not stop. Continuing data collection throughout the entire behavior program and into follow-up ensures ongoing evaluation to determine if the program is effective. Collecting a sufficient amount of data proves to be a challenge on a regular basis for behavior analysts working in the field, because they often do not have the resources to do good online evaluations. Sometimes behavior analysts may be swayed by the clients' claims that they are getting better and that *their problem* is going away, but these anecdotes are no substitute for reliable, objective data. As a behavior analyst, you can argue that you need to take the time and trouble to continue to evaluate your interventions because this is required by our code of ethics (Guideline 4.04). A closely related issue has to do with making modifications to a behavior program. Here the guidelines require that any change be based on data (Guideline 4.05). Staying on top of each and every program you write not only is an essential strategy for effective behavior analysts but also is required by our Guidelines for Responsible Conduct.

TERMINATING CLIENTS

One issue that comes up repeatedly in our ethics workshops has to do with how ethical behavior analysts terminate a case. It is, of course, unethical to just drop a client (Guideline 2.15). The most ethical thing to do is transfer the client to another professional if you cannot continue with the case. If you have completed your objectives with a client, it is absolutely appropriate to terminate services. This, however, should not come as a surprise to the individual. It should be covered in a plan for "appropriate pretermination services" (Guideline 2.15d).

One exception to this general rule is "client conduct," where it is assumed there has been some dramatic change in the status of the relationship that would make continuing dangerous. A recent case was described to us that involved a single mother of a developmentally disabled child. The child was receiving behavioral services in the home. The young mother acquired a live-in boyfriend who, apparently, was a drug dealer. When the behavior analyst came to the house, the mom was high, the boyfriend had locked himself in the bedroom and was yelling threats through the door to anyone within earshot, and there was evidence of drug activity in the house. The behavior analyst left immediately and informed her supervisor that she would not go back to the home. Understanding her obligation to her client and "to all parties" (Guideline 2.02), she recommended that social services investigate the situation with an eye toward removing the child from the home.

DEALING WITH COLLEAGUES

Professional behavior analysts work with colleagues from two categories: other behavior analysts, and people from other allied professions (doctors, nurses, teachers, social workers, physical therapists, etc.).

Behavioral Colleagues

As unusual as it seems, from time to time, colleagues who are behavior analysts will present challenges. For example, some behavior analysts work for an agency where there are 10 or more Board Certified Behavior Analysts® (BCBA) or Board Certified Assistant Behavior Analysts®. Nowadays, these behavior analysts can have completely different backgrounds in terms of their training. Some of them may have taken their course work and practicum in a standard 2- to 3-year graduate program with faculty specially prepared for the task. Others may have taken the bulk of their work online and then received their supervision hours later. The exposure to ethics training for both of these groups of behavior analysts could range from a few lectures to an entire 3-credit-hour course over a standard 15-week semester. If you were a graduate student who sweated in three classes a week for 15 weeks over scenario after scenario trying to figure out the most ethical solution to a complex problem and then you had to defend your answer in front of your colleagues, you may be more sensitive to ethical dilemmas than a student who just listened to a lecturer on the computer. In any event, you should know that everyone does not take the task of worrying about ethical conduct as seriously as you might.

If you carefully study the Guidelines for Responsible Conduct, you will notice the guidelines cover the most frequently occurring situations you will encounter. If you work in the vicinity of other behavior analysts, you may see occasional violations of the guidelines, and you'll need to decide whether to take action. Fortunately, Guideline 9.01 can provide you with some assistance, because it directs you to "attempt to resolve the issue by bringing it to the attention of that individual" (Bailey & Burch, 2005, p. 249). Where you draw the line on the ethical violation is not clearly spelled out, but the prior commitment to integrity (Guideline 1.05) suggests that it would not take a major infraction for you to respond. Because you may be dealing with a fellow employee, it is best to start with a simple, unaccusatory question about what you know or saw. As per Guideline 9.01, ask for a meeting, explain your concern, and look for a reaction. There is a good chance the person's response will satisfy you, and you can move on with your life. If you still have a concern that the person is presenting a threat to the client, the company, or the profession, you will need to let your conscience be your guide, because the guidelines do not specify further action on your part. Sometimes it will be a good idea to seek advice from your supervisor or the supervisor of the other person.

In a recent case, a BCBA thought she saw what appeared to be a conflict of interest between a colleague and a single parent. The appearance was that the behavior analyst (female) had more than just a professional relationship with a client (newly divorced male with custody of an autistic child). The issue was raised with the colleague, and she immediately took action to transfer the child client to another BCBA.

In another case, a colleague described how he received two bottles of expensive wine from a client who was so appreciative of his consultation in a due process hearing with the school district. In an informal, nonthreatening manner, the staff raised the issue of the slippery slope of receiving gifts from clients. A lively debate ensued with no resolution, and no further action was taken.

In another case, a BCBA observed a colleague entering fabricated data onto a data sheet (she knew the data was fabricated because it was for the day before and the colleague was out that day). The BCBA met with the colleague and addressed the issue directly. She described the response as "disturbing." "It happens all the time," he said. "In order for us to bill the hours, we have to show the data. The client was here, and I know what he did. It's just that no one wrote it down. I don't see what the big deal is." The BCBA reported that it was the "no big deal" comment that disturbed her most, and she subsequently began observing the whole operation more closely. She soon concluded this was a practice that was common in this agency. Also common was the agency's providing services to clients who were not entitled to the consultations. The services were then billed to other clients who had more approved hours, "in order to help more people." The BCBA brought this to the attention of her supervisor and was given a lecture about being a "team player" and "working for the greater good." A few days later, this ethical behavior analyst put in her 30-day notice and took another job in another state for less pay.

Nonbehavioral Colleagues

Dealing with the ethical conduct of nonbehavioral colleagues presents a different sort of challenge because they are not required to subscribe to *our* guidelines. And there is a very good chance that their profession does not have anywhere near the commitment that we do to data-based decision making. The best advice

> "Dealing with the ethical conduct of nonbehavioral colleagues presents a different sort of challenge because they are not required to subscribe to *our* guidelines."

is to tread lightly when dealing with these colleagues and try to establish a working relationship (Guideline 2.03) in the best interest of your client. If you have established yourself as a reinforcer for this nonbehavioral colleague, and if you have demonstrated that you are a person of integrity, you may be able to ask questions or begin a discussion about some actions you think are unethical. Choose your words carefully, be sure to start with some open-ended questions, and avoid direct accusations. It is possible that you are wrong or that you have misinterpreted some action. You do not want to offend this person. Ideally, you'll be able to bring her to see the situation the way you do. Your hope is that the person will change her decision or take a different path toward effective treatment.

The ethical dilemma "Anonymous" presented at the beginning of this chapter is a difficult one, but it must be resolved by some direct action on the part of the behavior analyst. He has an ethical responsibility to resolve the conflict (see Guideline 6.06) and will need to deal directly with the administrator. We generally advise behavior analysts to take a copy of the guidelines with them to the meeting and point out the particular items that are relevant. In this case, in addition to highlighting Guideline 6.06 ("clarify the nature of the conflict … seek to resolve it") the behavior analyst could point out Guideline 1.05 (integrity) and the better part of Guideline 2.0 (responsibility to clients). Most facilities receive state or federal funding and are subject to review on a regular basis. Knowingly operating on false information is a form of fraud that is punishable by fines and revocation of a facility's permit. A young, somewhat timid behavior analyst may not be willing to push the issue this far but clearly could and would be right to do so. The code of ethics for our field was specifically designed to provide the support for professionals in the field who wish to protect clients' right to effective treatment.*

SUMMARY

In the course of their daily work, behavior analysts must adhere to the Guidelines for Responsible Conduct published by the Behavior Analyst Certification Board. It is also important that behavior analysts have integrity and uphold high moral standards. People who have integrity don't promise what they can't deliver. They cannot be persuaded to take a different position simply because it is popular or will bring them favor with others. Adhering to the guidelines, behavior analysts with good ethics will recognize that our responsibility is to all parties affected by our services and that as a field, our commitment is to evidence-based treatments. Above all else, we will "do no harm." Maintaining good ethics in both our professional life and our daily life will ensure that we are regarded as truthful, reliable, trustworthy professionals.

FOR FURTHER READING

Bailey, J. S., & Burch, M. R. (2005). *Ethics for behavior analysts: A practical guide to the Behavior Analyst Certification Board Guidelines for Responsible Conduct.* Mahwah, NJ: Lawrence Erlbaum Associates, Inc.

Lattal, A. D., & Clark, R. W. (2005). *Ethics at work.* Atlanta, GA: Performance Management.

---

* Six months after this incident, the behavior analyst took another job in a different state.

Two

Basic Consulting Repertoire

8

Interpersonal Communications

At an intake meeting for an elementary school student who needed behavioral services, Jason, the behavior analyst, was overly direct about how he wanted the program to be implemented, the data he wanted the parents to take, and the requirements for the teachers and classroom aides.

School staff looked at each other as he talked, and the child's mother pushed her chair back from the table. Jason didn't notice when the principal tried to politely interrupt him, and he kept on talking. Totally insensitive to the reactions of the people around him, Jason dug himself deeper and deeper into trouble.

After the meeting, the principal called Jason's supervisor. The supervisor would have to sit in on several meetings to undo the damage Jason had done with his less than stellar interpersonal communications.

Once upon a time, behavior analysts were thought of as cold, unfeeling people who cared more about data than the humans whose lives they were trying to improve. Fortunately, those days are gone. We now recognize that having good, actually *excellent*, interpersonal communications is central to becoming an effective behavior analyst.

Throughout your working day, and in your off time as well, you will be interacting with your supervisor, colleagues, direct reports, friends, and, most especially, clients.* Each and every

---

* Your direct reports are those employees you supervise, who report directly to you, and who receive direction from you.

interaction with them provides an opportunity for you to learn something about them, for them to know you better, and for you to have some influence or leave a good impression. The basis of excellent communications skills begins with *likeability.*

WHAT MAKES A PERSON LIKEABLE?

Likeable people project a warm personality. These are people everyone wants to be around. Likeable people are friendly and empathetic. They like to laugh and tell stories. They are real, nonphony, nonthreatening people who smile a lot, and they are genuinely interested in others. Likeable people do not appear to be pushy, and they are accepting and forgiving.

People who are rigid, uptight, intense, opinionated, judgmental, and brittle are generally not liked, and they do not do well in the consulting world. If you have any uncertainty about how you come across to other people, ask a friend for a frank and candid assessment. Ask for a list of specific behaviors that you can target for improvement. Take the feedback seriously, and set yourself up for a self-improvement plan.

INTERPERSONAL SKILLS FOR BEHAVIOR ANALYSIS SETTINGS

Behavior analysis is a technology of behavior change based on science. The science is certainly the foundation for what we do, but maximum effectiveness of the technology depends on the interpersonal communications of its practitioners.

"Maximum effectiveness of the technology depends on the interpersonal communications of its practitioners."

If you are a beginning behavior analyst, the most common situations for using interpersonal skills will be related to your role in implementing and managing a behavior program in therapy settings or a consulting plan in business settings. When behavior analysts take on a new client (which may be an individual child, agency, or business), they go through seven stages that range from initially accepting the client all the way to terminating the services. In each of these seven stages, interpersonal communication skills will be needed.

Interpersonal Communications With Clients

Initial Intake With the Client The process of taking on a new case begins with an intake interview or meeting. At the meeting, someone will request behavioral services for a client. The request could have been made earlier, and the initial meeting will be to establish the services. The person who requested the services could be the case manager, parent, teacher, principal, school counselor, or vice president of human resources.

This meeting is extremely important for you because it establishes your position in relation to the client. Your goal is to quickly build good rapport (Bixler & Dugan, 2001, pp. 71–80). To gain trust, you must establish that you respect the client. You should be a good listener, show confidence in your behavioral approach, display a caring attitude, and have a friendly demeanor. You should maintain good eye contact with each person in the meeting. Read the body language of the client, the client's family members (or relevant people in a business setting), and the others at the table (Bixler & Dugan, 2001, pp. 81–89). If necessary, steer the conversation pleasantly back to the topic if people start to talk for long periods of time about off-task topics. Before the meeting begins, a few responses to "Did you go to the football game this weekend?" are fine, but a 30-minute discussion that delays the start of the meeting is not appropriate. Make sure you have done your homework so you can handle questions with ease. It's a big job, no doubt about it. You'll need to use your assertiveness (Chapter 2) and leadership (Chapter 3) skills and establish your solid foundation in ethics by demonstrating your integrity. It is a good idea to review some of your basic social skills before you proceed. Smile, use the person's name in conversation, and be a good listener.

Present Your Analysis   Following the intake, you will engage in a systematic review of all information related to the case, look into the published research on the topic, and do some informal observations. Next, you'll conduct a functional assessment to determine causal variables related to the behavior that has been referred. In some cases, this might take a few days or as long as 3 or 4 weeks. Your next step is a meeting with the client, client surrogate, department head, vice president of human resources, or other appropriate people. You should have been in touch with these people by phone or e-mail at least once per week (or on some other prearranged time schedule at their preference). In the meantime, continue to develop rapport, and let the relevant parties know that you are working hard for them.

The initial face-to-face meeting is one of the most important interactions you will have with your client. This is when first impressions count, and you can leave others with a great impression of you, or you can leave them sitting at the table wondering how fast they can have you replaced.

At this meeting, start casually, and put people at ease. Now it's time for the presentation of your findings. On the basis of your informal observations, functional assessment, and review of the folder, present your recommendations. This is when you must follow Guideline 1.06 from the Guidelines for Responsible Conduct, speaking in plain nontechnical language. Good eye contact is essential, and a firm, strong voice shows your client that you know what you are doing. Make a convincing case for your recommendations by using tips from Chapter 9 ("Persuasion and Influence"), and be prepared to negotiate if necessary (see Chapter 10).

Remember that the logic you are using in making your behavioral proposal is often at odds with what people think about how to approach human behavior. For

> "Remember that the logic you are using in making your behavioral proposal is often at odds with what people think about how to approach human behavior."

this reason, you will probably have to build your case slowly, watching for telltale signs that others are not following you or disagree. Sure signs that people just aren't buying what you have to say include their breaking off eye contact with you, shifting in their chair, and actually pushing back from the table. If you notice someone seems to be having a problem with what you are saying, it's best to immediately deal with these signs of disagreement by saying, "Let's pause for a second and see if you have any concerns." Do not point out the behaviors that signal they don't agree with you. Saying, "I can tell from the fact that you rolled your eyes and I heard you swear under your breath that you might have some concerns" will embarrass some people and won't help your case.

If you have data to show, make them easy to read and visually attractive. Remember that most nonbehavioral people might not be very good at reading charts and graphs, so these should be compelling rather than confusing.

Keep the goal in mind throughout this meeting that you want "buy in" from your client or client surrogate. If you use the technique of starting with premises that the client currently follows and build gradually on that foundation, you should be successful. If the client's beginning premise is an alternative therapy that is not grounded in science, you will probably have to explain your position, answer questions, and handle objections along the way. For a good starting point, note that one thing everyone can usually agree on is that everyone wants the client to improve (or a business problem to be solved). Begin with that, and work your way into introducing your behavior analytic approach.

One useful strategy is to make ample use of anecdotes and stories from your own experience. If the stories involve other clients, make sure you don't violate confidentially. Most people who are not behavior analysts

> "Most people who are not behavior analysts respond to stories, anecdotes, and analogies better than to data."

respond to stories, anecdotes, and analogies better than to data, and often it will help if you can paint a picture of a case similar to this one that you handled successfully. Such stories also show that you are a person whom people can talk to rather than a "behavioral techno-nerd" who wants to control their child or take over their company.

If there are no questions or significant concerns, you will ask your client to sign a treatment plan (or in some cases a business contract). Be sure to have any necessary paperwork ready. There will be some cases for which concerns expressed at the meeting will require you to go back to the drawing board. In these cases, the formal signing of the treatment plan or business document will have to wait a few days until the necessary corrections have been completed.

Present Treatment Plan for Approval   If you have done a good job of establishing rapport with your client so she has a strong trust in you, the presentation of the behavior plan for final approval should be low-key and rather short. Using all of your relevant Dale Carnegie skills, such as being a good listener and helping the client achieve her goals; your next step is to firm up an agreement for your proposal. You should make an explicit request for client consent and cooperation (Guideline 4.01). Depending on the circumstances, you may want the client or client surrogate to actually sign a copy of the treatment plan. In therapy settings, a whole team of professionals will sign the plan, indicating their consent.

Preparation and Training of the Mediator   In behavior analysis, much of the heavy lifting is done by someone who is close to the client. This person might be a parent, teacher, or paraprofessional (see Chapter 6). These people are sometimes referred to as "mediators," and there is a good chance that as a consultant you will be responsible for training them.

> "In behavior analysis, much of the heavy lifting is done by someone who is close to the client."

In business settings, you will train supervisors or managers to analyze behavior and implement new contingencies in the work environment. This can be quite a challenge, and your interpersonal skills will need to be very sharp. The first thing you need to recognize is that getting adults to change their behavior is quite difficult. Adults are often resistant to doing things in a new way, and many don't like to be told what to do. Some adults are uncertain and lack confidence in their new behaviors, and they may try to resist or show a change of heart in the effort. Although it is not often looked at this way, being able to break a task down into smaller elements so that it is easily taught (the tried-and-true "task analysis") is actually an important interpersonal skill that each behavior analyst must have in her repertoire. Explaining how you want something done, such as giving instructions for completing an assignment, requires patience on your part and an ability to clearly model the behavior.

Using generous amounts of positive feedback and approval is another interpersonal skill that is essential at this stage of the process. If you don't have much experience in training adults, you might want to approach your boss or supervisor and point this out so you don't take on your first assignment without proper preparation. To ensure that your mediator is up to the task of prompting behavior, delivering reinforcers, and managing contingencies consistently, set up several observations and role-play practice situations before transitioning the mediator to a new role.* Having trainees observe you carry out the task is critical for them to acquire the skill quickly and to give them a basis for questions they may have. As a training technique for adults, role-playing gives *them* confidence and *you* a chance to reinforce copiously.

---

* In school settings, when dealing with senior teachers who did not request help but were referred by the school principal, you might have to adapt to the circumstances to avoid offending those individuals who believe they are competent and don't need to change their behavior because "it is the child's problem," not theirs.

Pay attention to the *body language* of the people you're training, and watch for signs that they are getting tired or bored. Be prepared to change tasks, give them a break, or quit early. You may not be able to train your mediator in one session, so allow time in your timeline for two to three sessions or days. Taking data during the training and showing these data to the people being trained is a good method for providing motivation for them to master the tasks. Considerable judgment, however, needs to be applied in these circumstances. Consult with your supervisor if you feel you are getting "push back" from the teacher or other mediator.

Showtime: Interpersonal Skills When the Intervention Is Finally in Place When the day comes for the intervention to begin, a new set of interpersonal skills is required. Now, you must watch closely to make sure that your newly trained mediator is in fact following the protocol that you described in the plan. This is the protocol that you modeled, role-played, and practiced so hard to accomplish. When you are observing a newly trained mediator who is demonstrating a procedure, look for wooden performances rather than genuine interactions on his part. This type of performance is common in the beginning, but you can begin to shape on this by prompting the right behavior.

The ability to use descriptive reinforcement, such as pointing out to a teacher that his student is now actually more on task or to a supervisor that an employee is using safety equipment more frequently, is a major interpersonal skill for the behavior analyst.

> "The ability to use descriptive reinforcement … is a major interpersonal skill for the behavior analyst."

You will also need to be prepared for *troubleshooting* along the way. It often happens that we can't quite predict the outcome of an intervention and need to make changes in the plan on the fly. If changes are minor you can usually handle them on the spot by saying, "I'm sorry, I didn't think about this

before, but let me show you another way to give these prompts."*
Admitting that you made a mistake or forgot to include some detail
is OK and shows the client that you are human after all.

One final interpersonal skill is that of *shaping on the performance of the mediator.* Your job as the behavior analyst is to monitor everything closely and provide reinforcers to help give the
mediator confidence and strength. It is also important to begin
raising the bar very gradually so that more and more is required
to earn your approval. In addition to shaping, you will thin out
the schedule of reinforcement so that the mediator is eventually
brought into contact with natural consequences.

Online Monitoring, Evaluation, and Maintenance If you have successfully done the first five steps, you're ready for the next one.
It is now when you begin to gradually phase yourself out while
continuing to evaluate the mediator's performance and the effect
of the intervention.

In this maintenance phase, you should drop in occasionally to
check on the case, provide some feedback, review the data with
the mediator, and make sure that any data are properly logged
and recorded. Generally speaking, no new interpersonal skills are
needed here. Your job is to provide encouragement, let the mediator
know how proud you are of her, and begin thinking of how you will
ultimately count this case as a success as you terminate your involvement. Your comments will move from "*You* are doing such a great
job" to "What a great improvement I can see. *What do you think?*"
Giving the mediator the credit for the success of the project is the
most important skill that you will need to exercise in this phase.

Termination For the purpose of discussing interpersonal communications skills, we will assume for the moment that the
case you took several months ago has gone well; that you were

---

* You need to determine in your overall analysis of the setting how much control the
administrator or principal desires regarding behavior-change plans. Some people want
to be kept in the loop and receive periodic updates; others want to approve any small
change. You need to know this in advance.

successful in your search for functional variables in the classroom, home, clinic, or factory; and that you were successful in training the mediator (teacher, aide, parent, shift supervisor) on how to handle the referred problem.

Now you are ready to wrap things up. If you have gradually faded yourself out in the previous step, this should not be too difficult. Your mediator should no longer be dependent on you for praise and feedback and should be feeling good about his new skills in changing behavior. You can begin to think of this person as more of a colleague than a client, and letting him know that it is time

> "We do recommend some sort of symbolic end to the behavior-change process that includes a healthy dose of appreciation on your part for the hard work put in by the mediator to pull it off."

to move on is your only major challenge. Some consultants like to have some form of celebration such as a party or a lunch for the client when it seems appropriate to say good-bye. You will have to decide for yourself what the best method is in your setting for saying good-bye. A final wrap-up meeting may be enough, but there will be times when clients or staff members have overcome many obstacles, and something more (such as a group lunch) may be appropriate. However you decide to handle this, we do recommend some sort of symbolic end to the behavior-change process that includes a healthy dose of appreciation on your part for the hard work put in by the mediator to pull it off.

INTERPERSONAL COMMUNICATIONS WITH EVERYONE ELSE

Communicating With Your Boss or Supervisor

When communicating with your boss or supervisor, it is always a good idea to be careful in what you say. Do not discuss other employees, give away personal information in a weak moment, or let the person see you as a timid, paranoid, or threatened individual. Your primary form of communication with your supervisor will be in one-on-one meetings where you receive your instructions, receive feedback, and perhaps, if you are accepted as a colleague, brainstorm problems and search for solutions. Your goal in such situations is to be open, constructive, flexible, and creative. Most of all, you need to pay close attention to make sure you understand the problem and what your boss wants from you. In some cases, supervisors are looking not for action but rather for just a sounding board for ideas. In other situations, they want someone to take care of a problem and relieve them of a burden. The extent to which you can meet these needs is the extent to which you will be able to gain support and further your own professional ambitions.

Communicating With Colleagues

Rule 1: Do Not Gossip   No matter how much you think you can trust people in the work setting, remember that they cannot keep secrets, and anything you say about another employee will get back to the person, usually sooner than later. When people want to tell you some gossip, it is a dead giveaway that they enjoy discussing the personal details of someone else's life. If they are talking about someone else, they are likely to talk about you. Some people use gossip as a lure to get you to dish out something in return for the juicy dirt you just heard. *Don't do it*. You will regret it when the person who was the subject of the gossip begins to give you the cold shoulder.

If someone starts telling you a tasty rumor, cut off the conversation or change the subject. Saying, "I really don't want to hear that" is a direct way of handling gossip. A less direct method is to say, "Let's get back to Billy's behavior plan."

In the work setting, the less you hear of the intimate details about others, the better off you will be. Although it is not covered under "confidentiality" in the ethics code, gossip can have the same kind of devastating effect as information gained from a school record if spread around.

Knowing that other employees can be so interested in spreading information, you can use this tendency to do some good. If you had particularly good luck with a certain teacher, be sure to tell this to others. With the way people like to talk, there is a good chance it will get back to the teacher. The same goes for your boss or other employees. A statement such as "Jayne really helped me out yesterday. I was in a jam, and she dropped everything to help me out" will circulate around the office in a day or two, and the next thing you know, Jayne will be offering to help in other ways as well.

Gaining a reputation in your circle of colleagues as someone who talks kindly about others, who admires others for their dedication and courage, and who is ready to give credit to others is far better than being known as a rumor monger and town gossip.

Rule 2: Do Not Discuss Salary or Company Benefits The way in which a company determines how much individual employees get paid and the benefits they receive is a private matter between the employer and each employee. Arrangements are made based on a person's value to the company, and by maintaining confidentially, officers can keep the harsh edges of pay differences from causing unnecessary friction among employees. If your company has a written rule about discussing pay and benefits, you will be told when you sign on. If there is no written rule, know that discussing your salary or the salary of others with people who have no business knowing this information is absolutely not appropriate.

> "Getting too cozy with your colleagues can have drawbacks."

Rule 3: Be Wary of Dual Relationships Most companies want a sense of camaraderie among their employees, and some go so far as to encourage activities such as participation in a company softball team, bowling league, or charity walk-a-thon. The goal is to produce a group of people who help each other work toward a common company or organizational goal. Being friendly, cooperative, and kind to each other rather than carping and being critical is certainly a worthy objective. To be a successful behavioral consultant, you will need others in your group to support you, and you will need to reciprocate. However, getting too cozy with your colleagues can have drawbacks. Becoming drinking buddies or constant companions can have the negative effect of causing jealousy. Jealousy then leads to rumors about cronyism, favoritism, or suspected sexual relationships in the office. Interoffice sexual liaisons, of course, can have serious repercussions and are strongly advised against.

COMMUNICATING WITH DIRECT REPORTS

Many organizations employing behavior analysts are small operations where the table of the organization is fairly flat. But effective behavioral consultants are likely to gain responsibilities for supervising others soon enough.

How should you communicate with these individuals? Basically, treat them the way you would like to be treated: with respect. You may have

> "Try to be the supervisor you always wanted but never had!"

had some bad experiences with a supervisor who was vague in handing out instructions, who wouldn't listen to you, who talked down to you, or who never gave positive feedback but was quick to criticize. Now is the time to correct all that with the people you supervise: Try to be the supervisor you always wanted but never had! Fortunately, our basic principles of behavior come in handy in this circumstance. Make sure you have good stimulus control when describing goals and objectives, and use clear descriptions of desired behaviors. Be precise in stating timelines and deadlines. If you are seeking input from your supervisees, make sure they know that you will listen to and reinforce good ideas. If you run into someone who doesn't pay attention or doesn't follow through with instructions, analyze the situation before you start criticizing. Maybe you weren't clear in your instructions, or maybe you were vague about when you wanted something done or exactly how you wanted it carried out.

And, finally, don't forget what you know about reinforcement. Be sure to always describe the reinforcer for completing a task. At the very least, there should always be an implicit if not explicit understanding that your praise and social approval are always there for employees who work hard, who are creative, and who show themselves to be valuable members of the team.

SUMMARY

In addition to being technically proficient, behavior analysts who want to maximize their effectiveness will need to have outstanding interpersonal communications. To begin with, the consulting behavior analyst needs to be a likeable person. If the behavior analyst has some deficits in the likeability area, getting feedback from truthful friends and colleagues is the first step toward improving. Professionals with good interpersonal skills know how to build good rapport, show a caring attitude, and act friendly (but professional) toward others. They are well prepared in meetings, and they deliver content so that others can understand it, using nonbehavioral language if necessary. They are patient and give plenty of positive feedback. From the initial intake meeting when it is decided a client will receive behavioral services all the way to the termination of the program when the goals have been met, effective interpersonal communications will play a crucial part in the continued success of the consulting behavior analyst.

FOR FURTHER READING

Bixler, S., & Dugan, L. S. (2001). *5 steps to professional presence*. Avon, MA: Adams Media.

Carnegie, D. (1936). *How to win friends and influence people*. New York: Pocket Books.

Harvard Business School Press. (2004). *Face-to-face communications for clarity and impact*. Boston: Author.

9

Persuasion and Influence

When Chrissy's teacher referred Chrissy for a behavioral evalua-tion, she was as much concerned about Chrissy's mom as she was about this noncompliant 12-year-old with ADHD. Mom started the meeting by telling us she was certain Chrissy behaved the way she did just to irritate people: "She knows exactly what she's doing, and she knows better." I knew if I came on too strongly about the need for a behavior program complete with data col-lection at home, a baseline, and functional analysis, I would lose Mom's support. I needed to calmly introduce my position, by first acknowledging this mother's feelings. Only then could I influence her to try a behavioral approach for helping Chrissy.

Russ was the performance management consultant for a large cor-poration. Tough times in business, new competition that was taking away customers, and higher costs to run the company were result-ing in severe budget cuts and downsizing. In the midst of all this, Russ wanted to propose a new program that had the potential to gain new customers and generate revenue, and he proudly said,

> *Nobody could believe I sold my idea to the president and board of directors when they had been saying, "No new initiatives" for quite a while. I designed some materials so they were camera ready for printing—these made my new program look real. Then, I started my pitch by reminding everyone how successful my other programs were. I knew that the main objective of the board is to make money, and I had plenty of financials in the presentation. There were some questions and initial objections, but I was ready for them. Basically, I persuaded everyone to let me roll out my new program within 6 months.*

From classrooms to corporations, behavior analysts are in the business of persuasion and influence most of the time. What we offer in the way of treatment is counter to the natural tendencies of many parents, teachers, and supervisors. For this reason, in addition to using our technical skills such as functional assessment, we have to be able to persuade and influence others.

A child who is defiant and refuses to clean her room, go to bed on time, and get up when called in the morning presents a "power problem" to her parents, and the noncompliant behavior is commonly met with a forceful reaction. Unfortunately, the parent shouting, "I SAID, GET UP NOW! I AM GETTING SICK OF THIS! YOU ARE GOING BACK ON RESTRICTION!" is a more typical reaction than the parent backing off and conducting a functional analysis to determine *why* the child is refusing to get out of bed in the morning.

Although behavior analysis clearly offers a more rational, humane, and effective approach to behavior change, our approach is not intuitive to most people. At least in the short term, many people have been reinforced for making threats. This history of reinforcement can drive people's immediate reactions to problem behaviors. Another part of reinforcement history is that most people who are seeking help have become frustrated with their inability to manage others, whether it is their children, students, or employees. This frustration signals to people that they are about to lose control, and the resulting fear can result in irrational actions.

This is where influence and persuasion come in. Our job as behavior analysts is to bring these frustrated people back from the brink, put them in a calm and reflective mood, and talk them into trying a different approach. We have to change *their* behaviors, and their attitudes, before we can get to the point of changing the behavior of the child or employee in question.

Attitudes are largely verbal repertoires that are somewhat connected to the behaviors of the individuals, not necessarily in a cause–effect manner but perhaps indicative of how they might behave in certain situations. A frustrated parent on the verge of using corporal punishment with a child is going to need some persuading to back off, cool down, and listen to reason. As a behavior analyst, you will see that your first challenge is to find the right setting to talk to the frenzied person as you use comforting words and reassuring gestures. This will help you put a crying mom or cursing boss in a place where he or she will listen and consider what you have to say.

INFLUENCE

Influence involves changing the attitudes and behaviors of other people without using any force or show of power (Harvard Business School Press, 2005). When using influence as a tool, behavior analysts turn on their best active listening skills, do a little shaping of any comments that lean toward the direction they are headed, and use reflection to make sure they are remaining in close

> "Influence involves changing the attitudes and behaviors of other people without using any force or show of power."

touch: "So, Ms. Phillips, what I hear you saying is that your frustration is so great that sometimes you actually lose it and feel like you're a step away from hitting your daughter. This has got to be upsetting for you." Once the parent has vented the agitated, irritated, and angry verbal behavior, he or she may be able to discuss the problem more rationally. Only then will the behavior analyst be able to make any headway in bringing up a search for causal variables. In this particular case, you may have to teach a parent who is saying, "She's just a rotten kid, and I think she actually enjoys pulling my chain" to assist you in your search for antecedent conditions, motivating operations, reinforcement history, and current contingencies.

One goal of your initial interview with your potential client is to determine how to proceed with your presentation. It rarely happens that you are dealing with someone who is totally open to whatever you say. Before getting down to business and addressing the child's problem, you might have to deal with whatever the parent has heard about you, your company, or the field of behavior analysis.

In addition, you will have to determine the best approach for dealing with what you have just been told. Some people come to the table with a strong bias against any kind of reward, or they might be explicit about not wanting to confront their child. If some parents had their druthers, they would be perfectly happy

> "When put side by side with counseling, a behavioral approach sounds like a labor-intensive approach."

to head for the mall after dropping the child off at a talk-therapy counselor who could say a few magic words and pronounce the child "cured" after just a few sessions. When put side by side with counseling, a behavioral approach sounds like a labor-intensive approach to many parents. With some parents, you may have a *big* job ahead of you in the influence department.

TACTICS OF INFLUENCE

To influence the thinking of someone, such as the parent with the child who needs a behavior program at home, you will want to consider some accepted tactics of influence. These tactics include *framing the issue* your way, *using information for influence*, and *having influence through your technical expertise as a behavior analyst* (Harvard Business School Press, 2005).

Framing

Framing involves presenting the discussion in a format or context with which you feel comfortable. In the case presented earlier, the most likely frame would be "learned behavior." You are more likely to be successful in proposing a behavior program for this child, rather than simply condoning punishment, if you can change the mother's opinion by giving examples of similar cases you worked on where the behavior turned out to be learned and where you were able to analyze it and teach a more appropriate behavior. You should also include in the frame the notion of the parent taking back control, setting limits, and having rules and consequences.

Remember that influence involves a "no force" approach, and if you want to use influence when interacting with a parent, it is advisable to avoid getting into an argument about her unusual theories of child behavior. Although Dr. Phil sometimes responds to off-the-wall parent ideas with statements such as, "That's just nuts," this is not a good way to get buy in for your approach.

Information

A second strategy for having influence is to gather and use relevant factual information in your presentation. In the case presented earlier, in addition to the stories and anecdotes that you can offer to the parent, you can present easily understood graphic representations of behavior-change projects from your own work or from the published literature.

Many people try to rely on the presentation of data alone. This is a mistake. Many consumers either don't understand or don't trust data. They may even say, "Oh, I've heard anyone can lie with statistics. I just don't believe that." In this case you'll

> "Many consumers either don't understand or don't trust data."

want to back off in a hurry. Of course, having data is the strong suit of a behavior analyst. But for many people, a stronger selling point than your having data will be the fact that they trust you. If you say you can help the parents, you seem like an honest person, you have taken a personal interest in them and their child, and, most important, you come across as a caring and responsible professional, you will be able to persuade them to give you a chance.

Technical Expertise

One final tactic of influence is to use your technical expertise in behavior analysis to convince the decision maker that you truly understand the problem and that you are uniquely prepared by your training and experience to solve it. This tactic is not used in every case, but it may be relevant when the behavior is uncommon. An example of an uncommon behavior problem is an eating disorder, where the behavior is very serious or possibly life threatening. If you took an internship at the Kennedy Krieger Institute and worked in the Pediatric Feeding Disorders Program for a semester or worked under Dr. Brian Iwata in his lab for self-injurious behavior at the University of Florida, you could claim expertise in dealing with very serious behavior problems. This level of expertise greatly increases your ability to influence a parent to try a sophisticated strategy for behavior change.

PERSUASION

Persuasion is a popular topic, and it is often referred to as *gentle persuasion*. Authors have used the term *gentle persuasion* when talking about converting people to a religion, when referring to what the media does to us, and when talking about sex before marriage. Gentle persuasion is known as a method to train horses and it is the name of a Scottish folk album. Despite the popularity of this term, we believe that *gentle persuasion* is a somewhat redundant expression, because by definition, persuasion is always gentle. Otherwise, it would be called something like *manipulation* or *arm-twisting*.

For our purposes, we defined persuasion as a process for logically presenting your point of view to a group that is responsible for making an important decision. Persuasion differs from influence in that influence involves producing an effect without using force, and persuasion involves changing an opinion by presenting your view, sometimes by using argument or debate.

An example of when you would need to use persuasion is when you, as the behavior analyst for a school, are asked to present a plan to a school principal that might totally change the discipline policy and procedures. Or you would want to use persuasion if you were the performance management consultant assigned to an insurance company, and your job was to present to the CEO your company's plan for rolling out a new high-performance training program.

The four elements of persuasion are having credibility, understanding your audience, making a solid case, and effectively communicating (Harvard Business School Press, 2005).

Having Credibility

If you are going to take responsibility for a major effort in an organization or corporate setting, you have to start by determining if you have the credibility to do the job. With this major responsibility, you need the trust of both your company and the leaders of the organization. This sort of trust comes only after you gain considerable experience in various leadership positions. Before taking on the responsibility of making a major presentation or leading a large project, you will have worked yourself up through the organization and proved that you can accomplish objectives. This means that you have demonstrated an ability to bring projects in on time and on budget.

> "If you are going to take responsibility for a major effort in an organization or corporate setting, you have to start by determining if you have the credibility to do the job."

If you are sincere, honest, and reliable and have proved to your colleagues and the leadership that you have integrity, you will develop the credibility you need to be a persuasive individual. When you combine this with experience and proven effectiveness, you will create an image for yourself that you are a person who can be counted on. The formula for credibility,

Credibility = Trust + Experience
(Harvard Business School Press,
2005), is a good one to adopt as a
personal goal.

> **"Credibility = Trust + Experience."**

Understanding Your Audience

Another element of persuasion is understanding the professionals
you are talking to and how they typically make decisions. Gather
as much information as you can on the process they use, and
try to get some sense of who on the decision-making team is the
leader. Assess the leader's influence with the group, and analyze
any recent decisions that he has made. Remember that the saying
"the best predictor of future behavior is past behavior" applies to
groups as well as to individuals. Although you might like to think
that a decision on your proposal is going to be based solely on
its merits, the truth is that individual members of the decision-
making body may have their own agendas, they may be jockeying
for position on the committee, or they may try to use your proposal to reach some other, unstated, goal.

As you present your case, watch closely for reactions from the
committee members, assess their body language, and be prepared
to pause for questions or clarifications. In addition to considering
the various motives of the panel members, you also have to take
into account the extent to which they are even receptive to your
plan. The committee members' response can range from hostile
to supportive (Harvard Business School Press, 2005), and you'll
need to react accordingly.

Making a Solid Case

The heart of your persuasion effort is the case that you make for
your project. It should be logical, meet the needs of the company and
those who will be making the decision, and take into account the
politics of the organization (Harvard Business School Press, 2005).

Taking a tip from the Greeks, you can consider a five-part
approach to building your case: (a) start with a compelling story;

(b) go straight to the heart of your proposition by painting a picture (by showing some pictures or video clips, for example) of what you want to happen; (c) present your factual evidence, supporting it with easily read graphs, charts, or diagrams; (d) anticipate and answer any objections that are likely to come up; and (e) finally, end with a clear statement of what you want the panel or committee to do (Heinrichs, 2007).

Rhetorical devices that can be used persuasively in your presentation include asking rhetorical questions to get the decision makers' attention; varying the pace, pitch, and volume of your speaking voice; and using dramatic pauses to emphasize certain parts of your presentation.

Effectively Communicating

Clearly, the way in which you present your case is the key to persuasiveness. There is a variety of ways you can approach presenting a case. One way to begin is to start with something dramatic. A shocking statistic, a compelling story, or a humorous example of something gone horribly wrong (or beautifully right as the case may be) will get the attention of the committee. Once you have the committee's attention, follow up with your plan, and show how it will solve the problem. Continue the use of stories and metaphors to paint a picture of a better world if only your proposal were to be adopted. Another strategy is to start (after using some attention-getting hook) by analyzing a problem (much like we would do a functional analysis) and then presenting your plan as the solution. If you use this strategy, make sure that the solution does in fact match the dimensions and characteristics of the analysis. In either case, it is critical that you remember to focus your plan on how it will benefit the members of the committee individually and collectively. You can emphasize the improvements that will come if your approach is adopted: "We should be able to increase productivity by 5%, for an annualized return of over $3 million." Or you can talk about what will happen if it is not approved: "We will continue to lose market share to our international competition, and this could amount to more than $5 million over the next 5 years."

Behavior analysts often get so caught up in their data-based, evidence-centered strategy that they forget to take into account that most decisions are not made on logic alone. Decisions are made by people who have feelings, and you can appeal to this emotional side. If you started your presentation with a great story or humorous anecdote, you can follow it up with another example that reinforces the idea that your proposal will indeed solve the problem.

The words you use are extremely important, and when you write out your presentation, you should look for ways to add color and excitement. Painting a picture with words or accompanying your words with pictures, animations, or video clips can further sell your proposal. Saying things such as "Going to school is like going to work" or "Employees won't give their best if they aren't recognized. Do you think Olympic athletes would compete if there weren't any gold medals?" is a great way to frame your argument and get your audience to understand your point of view. Motivation is important to everyone.

Finally, to convince anyone of anything, you have to be able to tell a good story, one that is relevant, is rich in its details, exposes an emotional element that appeals to everyone, and has a conclusion that is satisfying and ties up all the loose ends.

> "To convince anyone of anything, you have to be able to tell a good story."

One behavior analyst was asked to do a presentation on her work with elementary school children. She initially started with charts and graphs that showed the prevalence of certain types of inappropriate behavior, and then she described categories of interventions based on their effectiveness. It was descriptive, it was accurate, and it was important, but it was lifeless.

She was asked, "Think back on some recent cases, and tell us a story of one child who stands out as a success." She thought a moment and then started to describe Lucy. Lucy was driving her teacher crazy, so much so that the teacher actually left the classroom and chased the behavior analyst down the hall, begging for help: "I don't know what to do. I can't sleep at night, and I'm seriously thinking of quitting this school." Now there is a compelling story, one that needs to be told in just the right way to persuade undergraduates to consider going into the profession as a behavioral consultant. The presenter was given the suggestion to start her presentation with this story, then proceed to describe what applied behavior analysis is and how it is used in the schools, and finish with the end of the story where Lucy became an A student and the teacher decided to continue teaching.

SUMMARY

As a good behavior analyst who is striving to be effective, you should work to perfect your influence abilities, your persuasion skills, and your presentation skills (see Chapter 11). Learning to influence people without coming across as heavy-handed can be learned only with practice and careful attention to the outcomes. Like influence, persuasion is an area that deserves further study by those who want to advance themselves and behavior analysis. We rely on other people to adopt and implement our well-researched behavioral procedures. Persuading them to use behavior analysis procedures and protocols is often dependent on our abilities to use the important skills of influence and persuasion.

FOR FURTHER READING

Beckwith, H. (1997). *Selling the invisible: A field guide to modern marketing*. New York: Warner Books.

Goldstein, J. J., Martin, S. J., & Cialdini, R. B. (2008). *Yes! 50 scientifically proven ways to be persuasive*. New York: Free Press.

Harvard Business School Press. (2005). *Power, influence, and persuasion*. Boston: Author.

Heinrichs, J. (2007). *Thank you for arguing: What Aristotle, Lincoln, and Homer Simpson can teach us about the art of persuasion*. New York: Random House.

10

Negotiation and Lobbying

When I first started working in behavior analysis, I would go into a meeting determined to have my way no matter what. It was "my way or the highway" as they say. Only when I had more experience did I realize that my goal when negotiating in a meeting was not only to get what I wanted but to help the other side get something too. Win–win means both sides gain something, and it goes a long way toward building your reputation and helping you in the future.

A nationally recognized behavior analyst

Behavior analysts are in the *business* of providing quality behavioral services. In therapy settings, these quality services are composed of the evidence-based treatments and interventions that, when applied properly, result in dramatic changes in behavior.

In the course of a behavior analyst's work, there will be many times when someone needs to be convinced that a certain treatment approach is the best one or that a new program should be implemented. When you find yourself in the position of being the consultant who has to convince others a specific approach is the best one, you'll need skills in negotiating and lobbying, in addition to your technical skills, to maximize your effectiveness.

Negotiating and lobbying will also be relevant when the time comes for you to request a higher salary or better working conditions, when you need to add staff or resources to your program, or when you want to take a different approach to how a school or agency handles a big issue such as schoolwide discipline.

NEGOTIATION

There are a number of books on negotiation and dozens of models for negotiating anything, from increasing your salary to getting a divorce, reducing your credit card rates, purchasing a used car, and getting the job of your dreams. Authors have described different critical steps for negotiation, depending on the setting, that range from not getting caught up with science (not such good advice for our field) to assuming control of the meeting the minute you walk in the room. Although "seizing control" might be the kind of advice given to a high-level corporate attorney who will be attempting to close a merger and acquisitions deal, it is not the best approach for behavior analysts who are part of a treatment team.

For behavior analysts who will be negotiating in a meeting, we offer the following guidelines.

Negotiation: The Behavioral Way

Premeeting Behaviors

Identify Your Goals What do you want to get out of this meeting? Do you want data to be collected every hour? Do you want a one-on-one behavioral assistant assigned to your client? Do you want the administrator to allow you to take a baseline before you start a new facility-wide program?

Do Your Homework Know who will be at the meeting and what positions they are likely to take. Is another professional going to resist your approach? Are you getting ready to propose the purchasing of services that are not in the principal's budget? Is the teacher you are working with resistive to taking data because she feels she has too much to do? Know how the agency or company works. If you are dealing with people from a nonprofit group with a limited budget, you might find yourself embarrassed when you suggest they hire a full-time PhD-level behavior analyst. As you develop advanced consulting skills, a part of "doing your homework" may involve lobbying for what you want. Lobbying is discussed later in this chapter.

Come to the Meeting Prepared   If you are working with a child client, have the relevant data graphed and ready for a brief presentation. If you plan to suggest purchasing equipment, find out in advance what the costs and options are. If the topic is a child with a behavior problem, observe the child in advance, if possible, and talk to as many of the other team members as you can who have worked with the child.

In-the-Meeting Behaviors

Identify the In-Charge Person You don't need to announce this, of course, but you should look around the table and determine who has the ability to make decisions. So many times, during meetings and outside of meetings, people waste their time talking to the wrong person. For example, they'll talk and talk and talk to an executive director who has no authority at all and simply serves at the pleasure of the board of directors. If people want to be effective, they need to go to the organization's president or board. As a behavior analyst, you should interact with all of the members of a team, but you should keep in mind who has the final say on issues. There will be times when a better approach than your trying to work out something at the treatment team level is for the owner of the consulting firm for which you work to meet with a district administrator or company owner.

Present Your Position   When it is your turn, give a very brief summary of the situation, then state your position. For example,

> We're meeting today because Shawn has injured two other children in his class, and as a team we need to come up with a plan. I believe we should begin a functional analysis immediately, and until we know what is going on, Shawn needs one-on-one supervision.

Or,

> The behavior analysis team has been asked to develop a number of behavior plans for our adult clients. It is our feeling that the behavioral issues are a result of the clients having nothing to do for the hour before lunch. We would like to propose the facility implement a before-lunch leisure skills program.

Understand the Position of the "Other Side" In the meeting, when others say they don't agree with your proposal, they don't want to take data, they don't think a behavioral approach is needed, or they don't have the resources to do what you are suggesting (you get the picture), make sure you understand their position: "Debbie, you said it is not practical for us to take data in your classroom. Can you tell me what the issues are?"

Identify the Points on Which Both Sides Agree Put the problems aside momentarily, and talk about the points that you agree on: "Can we take a step back for a minute? Debbie, do I have it right that Shawn is hitting other children in your class and that we do need some kind of an intervention?"

Compromise When Possible Behavior analysts should consider themselves part of the treatment team. Remember, your goal is to maintain your high standards and follow the Guidelines for Responsible Conduct, but you should work with other members of the team. Sometimes this will mean that you help with observations, agree to take data less often (provided it will not affect the quality of your services), change a schedule to help out, change a location to help out, and so on.

"Remember, your goal is to maintain your high standards and follow the Guidelines for Responsible Conduct, but you should work with other members of the team."

Summarize the Negotiated Agreement   Someone in the meeting should document what the team members agreed to do. This summary should be done verbally and provided in written form (this can be in meeting minutes or in your own notes).

Know When to Walk   As a behavior analyst, you'll strive, we hope, to be a team player. But there will be times when you are asked to do something unethical, sign off on services that are simply inadequate, or implement a program that is not going to result in effective treatment. When this happens, your best ethical course of action is to refuse to provide services. If you are working for a consulting firm or you're a graduate student in a student placement, you should always seek guidance (that's what cell phones are for). Contact your supervisor, and discuss the situation to make sure you are on the right track.

THE NEED FOR NEGOTIATING AND LOBBYING

To further describe the behavior analyst's need for negotiating and lobbying skills, let's consider the situation where you're called in on a case for a child in a public school program.

In the process of interfacing with a mediator or an individualized education program or habilitation team, you might suddenly discover that your diligent research and development work is about to be compromised by someone, or a group of people, usually not trained in behavior analysis. For behavioral procedures to be effective, we, as a field, have evidence that certain conditions are necessary. For example, consistency (meaning no one is giving intermittent reinforcement for inappropriate behavior) is extremely important in behavior reduction programs. The practical implications here are that it is likely to be your job to convince a team that everyone, including the parents, teachers, and other professionals, needs to follow through with the behavior plan.

So, you've done your best work and come to the staffing only to discover that the committee is about to veto your plan or modify it significantly. "Scale down" is often a polite way of telling you that you are not going to get your way on this and that someone at the table does not believe the behavioral intervention is a critical part of treatment. In some cases, a "shortage of resources" may be the reason given for not allowing you to implement your proposed intervention. In other cases, the mediator may balk at the program because she just doesn't want to "spend that much time" on the training or contingency management you've so carefully designed. When this happens, you'll need to draw on your negotiating skills to get the best services for your client.

Where Does Behavior Analysis Fit In? Why Do We Have to Negotiate?

So, why doesn't everyone appreciate what we do? Let's back up for a minute and look at how behavior analysis fits into the larger system of human services. First, we have to admit that we came to the table much later than the other professions. Children have been presenting serious behavior problems in classrooms for years. For at least 30 years, there has been a documented literature on treating these problems, and that literature came out of education, not behavior analysis. Schools first responded to children who had special needs by developing "special classrooms." The early view on children with a behavior problem in a school setting was that it was the children's fault. To get these children out of the teacher's hair, a school psychologist would be brought in to test the children and send them on their disruptive way to a "resource room" down the hall or to a portable classroom out behind the main school building.

Another misguided model for addressing behavior problems in schools involved trying to talk to the children about their inappropriate behavior. Children who for some reason could not be tested out of the classroom (e.g., their IQs were too high) were sent to a school counselor. Often, the counseling was just watered down "talk therapy." Again, the theory was that there is something wrong with these children that needs fixing.

It is surprising that in some schools, removal to special classes and counseling are still the methods of choice for handling children with behavior problems. Although special classes and counseling are well established in the educational system as solutions, these interventions don't have a strong database to show that they actually work, except by removing the child from the original classroom, thereby immediately solving the problem for a very happy teacher. The obvious negative reinforcement effect in this model explains its popularity in school systems throughout the country. Any blame for the problem is shifted to the child, a powerless victim of a well-entrenched bureaucracy.

Now, since about 10 years ago, when there is a problem related to a child's behavior, along comes a consulting behavior analyst. Rather than testing the child, the behavior analyst goes into the classroom to observe what is going on and tries to find the causal variables for the child's disruptive behavior. What the behavior analyst discovers is that this student is not well served by the teacher, the material is too difficult for him, he is required to sit for long periods paying close attention to texts he finds uninteresting, and he gets attention ("get back in your seat") from the teacher only when he gets out up and walks around or disturbs another student. This child doesn't need testing and placement in another class, and he doesn't need counseling. He needs a new set of interesting academic tasks for which he can earn reinforcement. The reinforcement will come from a teacher who understands his need for praise. It all sounds wonderful, but be aware that this series of recommendations is not likely to make the behavior analyst very popular with teachers who are locked into the negative reinforcement model.

What's a good behavior analyst to do under these circumstances? First, you must recognize that a school is an entity with its own power structure, unwritten rules, and codes of conduct. A school is different from almost any other institution in our culture because the teacher has total autonomy in her classroom. She makes the rules, and she calls the shots. Even the principal, the parents, and the school board are reluctant to take any action when a long-time teacher is involved in a disagreement.

So, you're the behavior analyst working in this school. You've just received a referral from the principal to work with a disruptive child in Ms. Baker's class. You make the observations and draw the conclusions described earlier. What next? Tell Ms. Baker she has to change her behavior? Tell her she has to give up her yellowed notes and deal with the student in a completely different way from the way she has been teaching for the past 25 years? And tell her she needs to spend a lot of time reinforcing a child she has come to dislike? You can imagine how the conversation is going to go. For beginners, this might seem like an absolutely daunting task. If you successfully turn around a teacher who was not committed to a behavior analysis approach, the word will spread, and you'll have other people waiting in line for your services.

To get started, a quick course in "Negotiation and Lobbying 101" will help you make some headway on behalf of your clients.

LOBBYING

As most people know it, the term *lobbying* relates to organized attempts by special interest groups to influence legislators who are considering certain bills in Congress. The key features are the organized attempts and where they take place, which is outside of the actual deliberative chamber. The term *lobbying* originated from special interest group representatives cornering legislators in the lobby or hallway and trying to influence them outside of an official meeting.

We use the idea of lobbying to suggest a method by which you, the newcomer to the organization, might ethically work to influence the actions of others to have a positive outcome. The special interest you represent is the student who needs an advocate to improve his chances at succeeding in school. As the consulting behavior analyst, you aren't simply arguing that the teacher needs to cut the student some slack. Instead, you are recommending a very specific set of changes that you believe, on the basis of your review of the literature, your direct observations in the classroom, and your years of experience, will be the best way to help the child.

So how should you proceed? Knowing what you know about the power of the teacher to call the shots, you are going to have to do some lobbying on behalf of the child and yourself. You should do this *before* you ever actually sit down with the teacher at the meeting in which you present your proposal.

Establishing Yourself as a Reinforcer

Step one is to make yourself a reinforcer. Find some time that you can meet with the teacher to get acquainted. Using what you learned in Chapters 8 and 9, present yourself as a nonthreatening person who loves kids, who is awed by teachers who do the tough job with them each and every day, and who is sympathetic to their plight. This should be sincere. This will probably take two or three conversations in the teachers' lounge, cafeteria, hallway (lobby), or anywhere else but the classroom. Watch for anything that indicates what the teacher's reinforcers are, and see if you can provide them. For example, if you notice she has a bulletin board with a butterfly theme, you might bring her a book about butterflies that she can borrow. If you hear the teacher likes a certain kind of music, you might leave a flyer on an upcoming concert or a clipping from the paper about her favorite musician. Or you can offer to help do something with the class such as read to the students, assist with an upcoming assignment, or provide a copy of a computer program that they have been looking for. We need to note here that in keeping with good ethics, you should not be bringing the teacher gifts or presents.

Ask Questions to Get Information

Next, after getting to know the teacher a little better, you can start asking some questions after an observation session in her classroom: "I noticed that William gets antsy when you assign math problems. Can you tell me anything about that?" No matter what her reply is, the proper response from you is, "Thank you, Ms. Baker, that really helps me understand William. I appreciate it."

As you interact with the teacher, you will be able to get some feel of how best to pitch your program ideas to her. At the point that you finally sit down in a meeting, it would be very helpful if the teacher likes you, trusts you, and knows that you won't ask her to do anything she can't do. And you will know how this particular teacher feels about changing the way she does things in the classroom and changing her behavior.

Assess the Teacher's Response

Because your program is probably going to involve some aspect of academic rearrangement (a different set of assignments, different schedule, more reinforcement), you need to probe Ms. Baker to determine

> "Some people in this world are rigid. When working with them, you have your work cut out for you. The lobbying effort could take a while."

how she responds to these ideas. Some people in this world are rigid. When working with them, you have your work cut out for you. The lobbying effort could take a while. Don't rush this. If you move too fast with a person who does not like change, you'll probably get a rejection. You should know that once people have rejected an idea, it is difficult to get them to reconsider.

If you are working with a committee rather than an individual, what you need by way of lobbying is multiplied by the number of people on the committee. There is one thing that observers of the legislative scene have known for a long time and that is if you want your proposal to be received positively, it's a good idea to get to know the members individually, chat with them prior to your presentation if possible, and bounce some ideas off them to get a sense of their positions.

In recent years, actual lobbying in Congress has gotten a bad name. This is in part due to lobbyists who have gone way beyond just getting to know the legislators and educating them about certain bills. These unethical lobbyists have also offered bribes, kickbacks, golf trips to Scotland, and junkets to Hawaii. Clearly, behavior analysts who are lobbying on behalf of a child should keep lobbying efforts on the right side of ethics and the law. Understanding whom you are talking to so you can deal with any concerns ahead of time is a prudent strategy for being an effective professional.

TIPS ON NEGOTIATION

Negotiation is another business term that may seem out of place in the effective practice of behavior analysis. Most people who hear the term *negotiation* envision a mahogany table the size of Cincinnati with management representatives on one side and labor leaders on the other as they hammer out a 5-year contract for General Motors assembly-line workers. It's a great image, but that is not what we are talking about here. The more common definition of negotiation carries the sense of what we have in mind. Negotiation is a discussion that takes place to reach an agreement. From the quote at the beginning of this chapter, you get some sense of an all too common scenario that plays out every day across the United States. A behavior analyst has presented her case for a behavior program that will greatly improve the life of a client, child, student, or group of employees. And, despite the behavior analyst's best lobbying effort, the proposal has hit some significant resistance from one or more parties or even the mediator himself. She used all the tips we offered in Chapter 9 on persuasion, but heads are shaking and people are saying, "We don't have the time to do that sort of program," "It's going to take resources we don't have," or even "I don't feel comfortable making all his reinforcers, what do you call it, 'contingent.' I think he's going to pitch a fit right there in the grocery store, and then what am I going to do?" The smart behavior analyst (that would be you!) will have planned for this possibility and have a Plan B ready. It's time to negotiate.

First you need to dream big on your proposal: "Ask for more than you expect to get" (Dawson, 2001, p. 13). In our field, most behavior plans are scaleable, that is, they could involve treatment that covers most of the waking hours of the client or as little as an hour or two per day. Anything less is essentially worthless, and this establishes your nonnegotiable point. Lovaas's original work (Lovaas, 1987) showed that in about half of the cases, 40 hours per week of one-on-one, knee-to-knee, discrete-trial training with autistic children could produce amazing results over a 2-year period. The children in Lovaas's study seemed to have "recovered" from autism and could be placed in regular education classes where they could not be discriminated from "normal" students. Later research has shown that 20 hours per week of the same intensity of treatment plus some lesser intensive group work can achieve approximately the same results.

If you pitched a proposal for 30 hours of therapy per week to a child's parents, and they bargained you down to 20 hours per week, you could live with it. If you started your proposed schedule of treatment with 20 hours per week, however, and the parents attempted to bargain you down to 10 hours, which might be below the threshold of effectiveness, you would have to make a decision as to whether you should provide the services.

Another strategy that Dawson recommended (Dawson, 2001, pp. 34–36) is the Feel, Felt, Found formula, which seems quite appropriate for the personal approach we take with parents, teachers, and other caregivers. In this strategy, when you are negotiating, you use words to convey the concept: "I understand how you *feel*, other people have *felt* the same way, but do you know what we have *found*?" This method of avoiding confrontation on key issues and recognizing the other person's concerns can diffuse a difficult moment and allow you to make headway toward providing a quality behavior program.

SUMMARY

The field of negotiation has many, many other strategies, tactics, and gambits to offer professionals who are playing hardball to win a contract, sell their services, or complete real estate deals. Most of these other strategies are not appropriate for human services. It is important to realize, however, that in the course of providing an extremely valuable, high-tech treatment for clients in need, the well-trained behavior analyst should be aware of contemporary lobbying and negotiation (i.e., bargaining) strategies. These skills are clearly worthwhile if they result in higher quality services for children and adults with disabilities who might otherwise be left with minimal services that have zero or a negative impact.

FOR FURTHER READING

Dawson, R. (2001). *Secrets of power negotiating: Inside secrets from a master negotiator.* Franklin Lakes, NJ: Career Press.

Greenwood, M. (2006). *How to negotiate like a pro: 41 rules for resolving disputes.* New York: iUniverse.

Lovaas, O. I. (1987). Behavioral treatment and normal educational and intellectual functioning in young autistic children. *Journal of Consulting and Clinical Psychology, 55,* 3–9.

Oliver, D. (2004). *How to negotiate effectively.* Philadelphia: Kogan Page.

11

Public Speaking

The presentation is simple, balanced, and beautiful.

Garr Reynolds (2008)

Sweating profusely and her heart pounding so loudly the audience can just about hear it, the phobic public speaker knows that her life as she knows it will be over just as soon as she opens her mouth. Word salad will pour forth, she will humiliate and embarrass herself with her stammering and loss of words, and for all eternity everyone in the audience will be left with the distinct impression that the bumbling speaker is one stupid, incompetent fool.

At the top of the list, the fear of public speaking is the second-most common phobia in the United States, between arachnophobia (fear of spiders) in the number one spot and aerophobia (fear of flying) in third place. Afflicting an estimated 5.3 million people, "social phobias," including public speaking phobias, relate to the fear of being evaluated negatively in social situations.

If you are in the class of people who have a fear of public speaking, with symptoms ranging from stomach grinding to full-blown panic attacks, the good news is this is

"The ability to speak to groups is a critical skill for behavior analysts."

a skill that can be learned with a systematic approach. The ability to speak to groups is a critical skill for behavior analysts. Giving presentations to civic or professional groups is probably the best way to spread the word about behavior analysis. With practice and training, you'll get to the point where you can address a large audience who will sense your caring and sincerity. Your enthusiasm for behavior analysis will come though loud and clear, and you'll be able to clear up any misconceptions that others have about our field. Section 10.0 in the Guidelines for Responsible Conduct is the "Behavior Analyst's Ethical Responsibility to Society," and this section of our code of ethics stresses the importance of promoting our field. In addition to meeting our ethical responsibility, there are many jobs for professional behavior analysts that will require you to have public speaking skills.

NEED FOR PUBLIC SPEAKING SKILLS

Our interviews with behavioral consultants indicated that stand-up training was one of the most frequently used skills in their repertoire. They described situations where, in some cases, as often as three times a week, they needed to address a group of people to talk about behavior analysis theory or techniques. Sometimes the consultants were simply building acquaintance with commonly used terms or basic concepts; in other talks they were conducting full-fledged training seminars with direct-care staff, pep talks for teachers, briefings of middle-level administrators, or role-playing and practice sessions with line workers in industrial settings. Consultants often described having to prepare a talk on short notice. For example, the behavior analysis consultant in a performance management setting might be given a briefing manual on Wednesday along with the instructions to deliver a 1-hour talk on behavioral safety on the following Monday for open pit mining field supervisors.

Many people simply cannot imagine themselves standing in front of a group of strangers and giving a talk. Like the phobic public speaker at the beginning of this chapter, they begin to think of every disaster that might happen during a presentation: "What if I choke up and forget what I'm supposed to say? What if they don't like me? What if the projector doesn't work? What if they ask me a question, and I don't know the answer?" Know that this line of thinking is usually common for beginning public speakers. Once you are practiced and experienced, you might have mild nervousness before a talk, but the intense panic will be gone. To get you started on the road to becoming a great public speaker, we'd like to offer some helpful suggestions. For experienced, confident speakers who have given talks and would like to brush up on their skills, we have some tips on catching the next wave in technology and public speaking theory.

HOW TO GET STARTED

If you feel comfortable talking one-on-one with people, you're off to a good start. If you have the ability to think of giving a talk as having a conversation on a topic with which you are familiar and excited about, you'll be fine. To get started, you should create some opportunities to carry on short conversations first with small groups and gradually expand to larger ones. This could take a few weeks, and during that time you could read some of the several good books on public speaking (e.g., Gelb, 1988; Harvard Business School Press, 2007; Henderson & Henderson, 2007; Hoff, 1998).

If, on the other hand, you're not simply a beginner but a person who has an actual mild fear of public speaking, you can approach this as you would a phobia; the best-known treatment of fears in behavior therapy is in vivo desensitization. As a behavior analyst, you are probably familiar with this technique. Just as you can address a fear of heights by starting out by climbing up one flight of stairs, you can work on your public speaking with relaxation training. You would envision yourself giving a talk with a small group of friends, then relaxing, imagining another person in the group, relaxing again, and so on, gradually adding people (Martin & Pear, 2006). When you are not fearful under these circumstances, it's time to try an actual talk on a very familiar topic with a group of people you know well. Desensitization works by gradual exposure to the fear-producing stimulus

> "To maximize your effectiveness as a behavior analyst, you need to push yourself to the point where you can address most any audience on short notice and tell your story with confidence and enthusiasm."

so that the anxiety gradually goes on extinction; how long this takes depends on how often you practice and will vary from one person to the next. If it looks like you are not making any progress by yourself, you might consider finding a behavior therapist to help you. It is worth it to do whatever it takes to overcome your fear. To maximize your effectiveness as a behavior analyst, you need to push yourself to the point where you can address most any audience on short notice and tell your story with confidence and enthusiasm.

STANDARD TECHNIQUES FOR PUBLIC SPEAKING

Preparing Your Talk

There are two parts to every talk: (a) the *content* (this is your message) and (b) the *delivery.* Most people worry most about the delivery and what they are going to do with their hands, how they are going to project their voice, and how they will move around the room or the stage. But it is impossible to be a great public speaker without having a compelling message to deliver. You need to begin with a story that you feel comfortable and excited about telling. This should be a story you know so well you don't need a script.

Step 1: Identify the Key Points of Your Talk   You won't have time to cover an endless number of details, so start by outlining your presentation to determine what is reasonable to cover in the time allotted. Most speakers attempt to include too much content, and then they feel rushed to cover it all. Big mistake. You'll be better off if you pare down your talk to about half a dozen important points in a 30-minute talk. Once you have your key points, determine the order in which they will be presented. It is very helpful to think in terms of a "story arc" with a beginning, middle, and end.

Step 2: Finding the Hook   Another tip for developing the perfect presentation is to start with the following mandate: "I need to give my audience a reason to listen to me. Why should they care?" This is the question a great speaker will answer within the first 90 seconds of a presentation. You should start with a hook to get your audience interested. A hook can be a question to the audience ("How many of you have had to endure someone sitting next to you talking loudly on his cell phone?"), a humorous story, a well-chosen quote, some dramatic statistic, or a "look ahead" that will result in your audience wanting to hear more. An example of a beginning that will cause the audience to think, "What happened next?" is the following introduction used by one behavior analyst consultant: "Little 7-year-old Jenny stumbled to her desk, pulled out her notebook, and wrote 'Help Me' in large letters with a stubby crayon. Then she put her head down and began sobbing." This was a well-chosen, dramatic beginning for a presentation given to school officials. This single anecdote had everyone in the room paying close attention to the speaker, who was giving a talk on individualized education programs that could have otherwise been routine and boring.

Step 3: Presenting the Content   You've started with a great hook designed to get everyone's attention. Once you have the attention of the audience, you can proceed to the content of your presentation. This will be the bulk of the presentation, although you must resist the urge to put too much material into this middle section. Your goal in the content section of the talk is to acquaint the audience members with the topic, show them how it is relevant, and motivate them to learn more. Rather than have slide after slide of bullet points that you read aloud, you will want to focus on the highlights as a progression.* Make sure the material is presented in an organized fashion and you adequately explain the concepts without burying the audience in details.

Step 4: End With a BANG! For the ending of your presentation, come back to the teaser and finish that story or provide one final anecdote that will give your talk a memorable ending. This could be a dramatic photo on the screen that tells a story in and of itself, a final humorous anecdote, or a quote that wraps up the essence of your talk. There seems to be a *primacy* effect and a *recency* effect (Gelb, 1988) with audiences. If people are asked later about the talk, they are more likely to remember the beginning and the end rather than the middle. This is another good reason to present just a few key points and follow them up with a handout that you distribute at the end of your talk.

ROOM SETUP AND EQUIPMENT CHECK

Successful public speakers arrive at their room 30 minutes to an hour prior to their talk. They do this because they understand the importance of creating a comfortable environment for their conversation with the audience. There are a few things to know about the room in which you will be speaking. First, the room was not arranged by a person who gives public presentations. It was most likely set up by the hotel catering crew the night before. Hotels have a few standard setups; the crew will simply follow instructions, and there is a good chance the room will not be set up the

---

* If your talk has a lot of content, this is best handled with a handout. Make this available at the end of the presentation.

way you want.* Second, it is your prerogative to ask for changes to make certain the room suits *your* needs and requirements. Many ballrooms are set with a riser (a small stage that is 1 to 3 feet above the floor of the room) and a podium. This is appropriate for

> "It is your prerogative to ask for changes to make certain the room suits *your* needs and requirements."

a large audience, but for smaller groups you may want to ask that the podium be put on the floor so you are closer to the audience. Or, even better, let your conference planner know if you want to use a wireless mike and will not be using the podium. For public speakers who are beginners, you may want to deliver your presentation from a podium. Standing in one place and clinging to the podium for dear life can give you an extra sense of security. But if your goal is to be a world-class speaker, know that public speakers who are at the top of the game move away from the podium, which puts a physical barrier between them and their audience. Great speakers connect with their audiences.

Lighting and Temperature

If you will be showing a presentation on a screen, make sure you know how to control the lighting in the room before the presentation starts. That way, when you're ready for the lights to be dimmed, you can say, "Could someone dim the front lights for me? It's the third switch on the back wall." This creates the impression you are well in charge of the room, and it prevents losing time while people scurry around the room turning the wrong lights on and off. The same goes for the temperature; as the presenter, you're in charge. If the room is too warm, especially when people have been

---

* A best practice of experienced speakers is to send ahead to your conference planner a floor plan showing how you want the room set up. This will need to be done at least 2 weeks prior to your talk. You can also specify at this time your audio–visual needs, for example, a wireless mic, the size of the screen, and the type of projector. If you are going to bring music, you will need to tell your conference planner that you want to plug into the hotel or room sound system.

listening to presentations all day, the chances are greatly increased that your audience members will drift off to peacefully nap during your talk. If the room is freezing, attendees will spend all of their time in your enlightening presentation thinking about how cold they are; those who aren't hearty members of the Polar Bear Club will just get up and leave.

Audience Seating

There will be times when you are scheduled to give a talk in a room that is too large for your presentation. Your talk is planned for 50 participants, and you are in a ballroom with seating for 1,000. There will always be audience members who choose to sit in the very back of the room, even though they will be 2 miles away from the speaker and there are hundreds of empty seats. As the presenter, you are well within the range of acceptable behavior to ask people gently and with good humor to move forward. If you are doing a workshop for which people need desks and tables, and you are assigned to a room with a theater-style setup (chairs only), it is very appropriate for you to locate the conference planner and ask how she can help you solve this problem. It might be possible to have a crew come to rearrange chairs or add tables or to move to another room. You'll be able to solve major problems, however, only if you have followed the suggestion of showing up 30 minutes to an hour in advance to check your room.

Testing the Equipment

Microphones   Once you have someone working on changing the room to your satisfaction, it's time to check out the equipment. Start with the microphone to make sure it works. Ask someone to respond to your question, "Can you hear me in the back?" If you are using a wireless mic in a very large room, the setup crew should explain the limitations of the equipment to you before your presentation begins. For example, after being turned on, some wireless mics have a several-second delay before they are operational. Unaware of this delay, nervous speakers will often turn the mic on and say something and, when nothing happens, they turn the mic off, declaring that it doesn't work. Some mics will not work if the person talking is standing in front of or near the speakers. Practicing a few seconds with the equipment before your presentation begins will result in your being cool, calm, and well prepared.

Projection Equipment   Next, check out the projection equipment. Most people use some version of a slide show. Modern-day slide shows (that used to be actual slides projected via a carousel projector) are projected using software such as Microsoft PowerPoint or Apple Keynote and a laptop computer. To begin checking out the projection equipment, plug your laptop into the projector and make sure it works. Advance a few slides just to make sure everything is working. It's a good idea to have a slide on the screen when people enter the room; this slide should have the title of your talk, your name and affiliation, and the name of the group that invited you.

Music: Moving From Acceptable to World Class

Adding music to your presentation and room ambience will make everything a little more complicated, but music is unbeatable when it comes to setting the mood and warming up the environment as the audience files into the room. You can bring your own music on an MP3 player or cell phone. These can be preprogrammed to give just the effect you want for your talk and your audience. For example, quiet classical music will relax everyone, whereas New Orleans jazz will get your audience energized and excited. In workshops we have done all over the country, one of the most common feedback comments is "I loved the music!"

Meet and Greet

If you get to your room 1 hour ahead of time and sort out all the equipment problems, you will have about 15 to 20 minutes to spare. People will be coming in and getting their coffee and taking their seat. Take this opportunity to meet some of the people who have come to hear you speak, particularly those in the first few rows to whom you will be talking when you start your presentation.

Delivery

The first recommendation for a great delivery is practice, practice, practice; this comes from countless books on public speaking and Toastmasters International, the leading organization in the United States that teaches and promotes good public speaking.* Practice in front of a mirror, practice with a friend, and finally, if possible, practice in the room where you will be speaking. Taping your practice talks and reviewing them in the privacy of your office will enable you to objectively analyze how your presentation looks from the audience side. Videotapes will show you if your body language was stiff and might provide the feedback you need to try a different style of presenting, such as stepping away from the podium, using a wireless mic, or actually walking right up the center aisle and around the audience where you can see people up close and talk directly to them.†

The second recommended practice for a successful presentation, right after cleaning out your pockets of course, is to relax. This might seem very difficult given the stress associated with talking to large crowds, but it works. Just before you go on stage, and

> "Remember that you are having a conversation about a topic that you know well and that you are here just to tell your story."

while you are out of sight, take several deep breaths. Remember that you are having a conversation about a topic that you know well and that you are here just to tell your story. These people wouldn't have invited you if they didn't want to hear your story. You'll be fine.

---

* Go to www.toastmasters.org for a complete set of recommendations on public speaking and a way to find a group near you.
† You will need a wireless remote as well if you are going to go mobile and show slides at the same time.

The final recommendation for a good delivery is to make sure that you are in your conversational mode and that you feel comfortable changing your tone of voice, from soft to loud and back, and using pauses for effect. If you have looked at yourself on tape and know that you have a tendency toward a monotone voice, absolutely make sure that you avoid that like you would avoid a giant pothole in front of your house. In a normal conversation with people, you will laugh, make jokes, and tell stories. All this is perfectly acceptable in your presentations as long as you keep it clean and the jokes and stories are relevant to your message. At a recent conference in front of 500 people, one of the invited speakers started his talk like this: "I've been told that I'm a pretty dry speaker and that I should start my talk with a joke. OK, so here's the joke ..." Unfortunately the joke wasn't funny, it wasn't related to the topic, and no one laughed. He had a hard time recovering, and even though he was a recognized expert in his field, he surely did not have the impact he wanted. To his credit, he had a 20-page handout with all the key points and references for people to pick up at the podium when the talk was over.

CREATING AN EXCELLENT SLIDE SHOW

Very few people can hold an audience's attention very long without a slide show to illustrate their points and make their arguments. People have become conditioned by watching high-quality, fast-moving television to expect to see and hear at the same time. In many cases, the audiences who come to hear our presentations at conferences will have been so spoiled by the high production values of television and movies that listening to a talk is plain boring by comparison. Speakers have to realize this and compensate not only by improving the delivery of their talks but also by upgrading the quality of their slide shows. Those of us who are a certain age will remember when conference presentations were done with overhead projectors and transparencies. Oftentimes, the transparencies were nothing but page after page of notes that we typed in a small font on an old contraption called a typewriter. Ambitious presenters then began showing pictures and graphs during presentations, and they did this by using slides, slide trays, and carousel projectors. Fast forward to the arrival of a technology for delivering presentations that we all know about: using laptop computers and software such as PowerPoint. This is clearly an improvement over slide projectors and overhead projectors. Rather than continuing to present a page of typewritten text, we all learned the rules of PowerPoint. Don't overwhelm the audience with writing. There should be no more than six words per line and six lines of text per page. If you're feeling up-to-date because you know all about these guidelines for preparing a presentation, hold on! Times are changing faster than you can blink, and cutting-edge presenters have moved beyond the PowerPoint presentation style that we all once knew and loved.

There is currently a counterrevolution going on against standard, ubiquitous, *boring*, PowerPoint-type presentations. Described as a reaction to "death by PowerPoint," this movement is out to revolutionize presentations by starting over with a new goal. Rather than asking, "How much text can I squeeze onto a page?" the group of creative designers leading the new wave proposed, "Great slide presentations contain appropriate content, arranged in the most efficient, graceful manner without superfluous decoration. The presentation is simple, balanced, and beautiful" (Reynolds, 2008, p. 25). Reynolds went on to say, "Live talks enhanced by multimedia are about storytelling and have more in common with the art of documentary film than the reading of a paper document. Live talks today must tell a story enhanced by imagery and other forms of appropriate multimedia" (p. 25).

> "There is currently a counterrevolution going on against standard, ubiquitous, *boring*, PowerPoint-type presentations."

This new approach, called *presentationzen* by Reynolds, takes its meaning from clean and simple Japanese designs for products. In fact, Reynolds credits an experience he had many years ago when he was eating lunch out of a *bento* (a type of Japanese lunch box) in a Tokyo train station. He happened to see a businessman flipping through page after page of PowerPoint slides printed two to a page. They were crammed with headers and bullet points, which were in contrast to his *bento*, which was "beautifully efficient, well-designed … nothing superfluous" (Reynolds, 2008, p. 6). Reynolds concluded that technical presentations could be made simple and beautiful as well. Reynolds's book is a must-read for anyone assigned to give a presentation or who wants to improve her public speaking. The basic concepts include having a minimalist approach to text, putting very few words on a slide, and not using lists of bullet points (gasp!). Instead, use captivating photos or compelling graphics to tell your story. For a 20-minute talk, no more than 20 slides are recommended. To see an example of simple, elegant presentations, go to www.ted.com. The TED mission statement, "TED* is devoted to giving millions of knowledge-seekers around the globe direct access to the world's greatest thinkers and teachers," along with suggestions for joining TED conversations is on the Web page.

SUMMARY

When it's time for you to give a presentation, don't think of yourself as having to "give a talk." The alternative to "giving a speech" is to be "in the moment with your audience." If you have designed a great presentation, have taken a few minutes to get to know some of the audience members, have fully practiced your talk, and are prepared to totally focus on your audience and share your story for the next 30 minutes with total disregard for anything that happened yesterday or before (with no thoughts about where you are going for lunch or when you will catch your

---

* TED is "technology, entertainment, design."

plane), you will be in the moment and ready to enjoy this experience with this wonderful audience, right now. If you can do this, everyone will recognize your efforts and will want to listen to your every word.

FOR FURTHER READING

Atkinson, C. (2008). *Beyond bullet points.* Redmond, WA: Microsoft Press.

Duarte, N. (2008). *Slide:ology: The art and science of creating great presentations.* Sebastopol, CA: O'Reilly Media.

Gelb, M. J. (1988). *Present yourself: Transforming fear, knowing your audience, setting the stage, making them remember.* Rolling Hills Estates, CA: Jalmar Press.

Harvard Business School Press. (2007). *Giving presentations.* Boston: Author.

Henderson, J., & Henderson, R. (2007). *There's no such thing as public speaking.* New York: Prentice Hall.

Hoff, R. (1998). *I can see you naked.* Kansas City, MO: Andrews and McMeel.

Martin, G., & Pear, J. (2006). *Behavior modification: What it is and how to do it.* New York: Prentice Hall.

Reynolds, G. (2008). *Presentationzen: Simple ideas on presentation design and delivery.* Berkeley, CA: New Riders.

www. Ted.com (Technology, Entertainment, Design).

Three

Applying Your Behavioral Knowledge

12

Handling Difficult People

What do the grumpy secretary at the front desk, the obstreperous colleague on your intervention team, the defiant and sarcastic employee you have to supervise, a self-centered supervisor who does not listen, and a CEO who is threatened by any idea that is not his own have in common? Answer: They are all people who fall into the category of "difficult."

Anyone who is employed long enough will encounter people who are difficult to deal with. "Difficult," of course, means different things to different people. Difficult people are, well, they're difficult. They are difficult to deal with, and they make things difficult for the people around them. In the case of the behavior analysis consultant, a difficult person is someone who slows down or derails our attempt to effectively implement our behavior-change agenda. Our behavior-change agenda can range from an individual client behavior program to the massive overhaul of a large corporation.

You'll notice in this chapter that we use commonly recognized labels such as *lazy, defiant,* or *self-centered.* We do this because these words can immediately conjure up a vivid image for the reader. In the work setting, however, as a behavior analyst, you'll have to quickly move beyond the labels and any emotions you are feeling, put on your behavior analyst hat, and begin looking at *measurable behaviors* and their causes in order to work with a difficult person.

Difficult people have a wide variety of behaviors and responses that make them difficult. Opposing new ideas, resisting feedback, lying, misrepresenting who really did the work, being manipulative, undermining and sabotaging others, bullying, dramatizing every issue, complaining, not complying with deadlines and protocols, criticizing others and their work, arguing about everything, pointing out why any new suggestion will not work, actually causing problems so they can appear to be the hero who solves them, and refusing to help, as in "It's not my job," are just some of the characteristics that can be seen in difficult people. These are people who prevent the work setting from being a calm, organized, effective place. They can push good workers over the edge and make them angry, frustrated, or depressed. They can derail projects and kill morale.

The nature of the work relationship you have with the difficult individual or where the person is on the organizational chart has a great deal to do with how you will approach this problem. For example, if you are pitching an idea to the vice president of manufacturing for a new feedback and incentive system and encounter resistance, this person might be difficult in a way that is totally different from the way a colleague criticizes your choice of clothing or whether you trim your beard. If you have a Board Certified Assistant Behavior Analyst® working for you who is a constant complainer, turns in sloppy work, or is always late in submitting her billable hours, this presents a different kind of difficulty.

Difficult people in the workplace are found at all levels, and it is important for the behavior analyst consultant to be able to deal with coworkers ranging from volunteers all the way up to the CEO.

DEALING WITH VOLUNTEERS OR MEDIATORS

One hallmark of behavior analysis is that we work with others to accomplish our objective of bringing about socially significant behavior change for a fairly large number of clients. A behavior analyst working in an elementary school would not be satisfied with one or two successes per month; she wants to implement programs with dozens of children and have an impact that can be felt all the way up to the principal's office. To accomplish this, a behavior analyst will have to work with teachers, aides, counselors, bus drivers, and perhaps even the janitors or cafeteria workers. The nonbehavioral people whom you recruit to assist you with behavioral programming are referred to as "mediators." Parents fall into this category if you will rely on them to carry out a behavior plan.

As described in Chapter 6, behavior analysts must be experts in conducting functional analyses in any setting, with any client who is referred, with the goal of discovering the controlling variables. The functional assessment is the first step in solving any behavior problem. This assessment leads to the development of a proposal for a behavior program that will alter the school-age child's behavior. The underlying goal of the behavior plan is to help the child adapt to the classroom environment and become a happy, cooperative, and productive student. Once the functional variables are pinpointed, the consequences that will be manipulated in the behavior plan are identified. Then, a person is identified who can hold and deliver the consequences in a consistent fashion to shape target behaviors such as improving on-task behavior, completing assignments, reducing error rate, increasing creative writing, or eliminating bullying. In many behavior programs, the key person who will be involved with delivering consequences (including reinforcers) might be someone such as the bus driver or a secretary in the main office. To be effective, the behavior analyst has to figure out a way to get these people involved and to get them excited about the prospect of helping a child change his behavior. This is where understanding difficult people comes into play. There will be times when the bus driver says, "I have to drive. I don't have time to be messing around with kids." In a corporate setting, there will be supervisors who resist reinforcing employees because, for example, "They are *supposed* to do their jobs. I don't get special rewards for doing my job."

Remember what we said about the functional assessment being the first step in solving a behavior problem? This applies to difficult people too. You'll have to ask yourself about the cause of their behaviors. Are they not getting enough reinforcement? Have they not been properly trained, and thus they feel uncomfortable doing the procedure? Does helping you mean they have to do their job in addition to what you need them to do?

But, in keeping with our message in this book, you'll have to go beyond the technical behavior analysis skills (such as functional analysis) and understand the political climate and the relationship of all the players and use all of the 25 skills that are essential for professional consultants.

Thankfully, there is increased recognition in the business literature that focusing on *behavior* rather than on the personality of the individual is a more productive strategy (Harvard Business School Press, 2004). It is the behaviors of the people you need to support your behavioral

> "There is increased recognition in the business literature that focusing on *behavior* rather than on the personality of the individual is a more productive strategy."

efforts (such as a teacher, teacher aide, or administrative assistant) that will be your target for change. If your helpers are difficult in the sense that although they agreed to assist, their not following through, not being precise in their actions, or being unable to provide the quality or consistency needed means the behavioral programming will not be effective. Who would have ever thought that a bus driver or cafeteria worker could make or break a behavior program?

It should be said at the outset that helping a behavior analyst implement a program can be a daunting task. Asking a bus driver to use differential reinforcement with a child who is sitting quietly involves not only attending to this behavior but also acting in a timely fashion to reinforce it. And the behavior must be reinforced *without the mediator making any errors.* It doesn't matter that the bus driver was the one asking for help with her unruly riders; she probably had in mind something a little different, such as banning the little miscreants from the school transportation system altogether. Saying, "I think I have a solution that will work for you" sounds good initially, but when the bus driver learns that this means praising children who are sitting quietly, including those who cussed her out yesterday, it's a totally different matter. This difficult person is essentially saying, "I'm not sure I'm up to this" when she fails to follow through with the behavior plan. One solution is to make the task easier by providing a bus aide for the first couple of weeks to help kick off the behavioral intervention plan. And there will certainly have to be some time set aside for some practice, with plenty of reinforcement for engaging in these new behaviors while driving a 38-foot, 12-ton school bus that is fully loaded with screaming middle school students.

The fundamental issue in dealing with mediators or volunteers is that as the behavioral consultant, you are required to first "qualify" them in the sense that you determine that they are in fact capable of carrying out their part of the intervention reliably, with precision, and for the duration needed. Once you have done this, these people must be motivated to participate and then trained to be a member of the intervention team. One final step involves monitoring their performance and providing feedback in the event that there is any drift from the specifications laid out initially. If you have done everything correctly, then the feedback will be positive, and you should be able to gradually fade from the scene. That is to say, if, as a behavioral consultant, you have applied what you know about staff training and used your persuasion skills effectively (see Chapter 9), you will basically avoid having a difficult person on your hands. If you haven't properly qualified the person or your training was insufficient or if the person encounters some negative reaction or aversiveness along the way, you may find that you have to now deal with that difficult person.

One of the most persistent complaints by behavior analysts working in the homes of autistic children is that one of the parents is sabotaging the treatment plan. This, of course, can deal a fatal blow to your success with the child. The difficult parent may engage in passive-aggressive

> "Employing a Declaration of Professional Services ... should help in screening out parents who are not likely to cooperate."

behaviors of agreeing to but not actually following the plan, "forgetting" to run a training session, or arguing with you about the need for training at all: "Don't you think he will just grow out of this?" Employing a Declaration of Professional Services when you are at the point of deciding if you will take a case should help in screening out parents who are not likely to cooperate.* Once into the case, you may discover that it is not easy for some people to follow a specific routine day after day, and you may have some second thoughts from them. Or one of the parents may talk to another family and start asking about alternative but potentially dangerous treatments such as hyperbaric oxygen treatment or special diets that might seem easier than rigorous behavioral training regimes. The best solution for these scenarios is to anticipate the problems and have regular check-ins with the parents to make sure they are still on board and committed to the behavioral program. If difficult behaviors arise (such as noncompliance), you will need to reverse gears and start over from the beginning and make sure that this time there is a sufficient amount of feedback and reinforcement to maintain their support.

As a behavior analysis consultant, you will find there will be many times when you'll need to work with mediators and volunteers to provide a child with services across settings and

---

* A Declaration of Professional Services is a multipage document that spells out the role of the behavior analyst and the caregivers in the treatment of the child. Details are provided that make clear from the outset that full cooperation is needed for success and that explain exactly what the parents will be expected to do. All parties sign the declaration, which has contractual language, and the parents keep a copy for future reference.

throughout the day. As a behavioral consultant, you'll also need the skills to supervise paid employees who will report to you in the work setting.

DEALING WITH DIRECT REPORTS

The people you supervise are referred to in the business world as "direct reports," that is, those people who report to you and have their performance evaluations done by you. Theoretically, they do what they are told, but this is not always the case. They can be lazy or argumentative, defensive, passive-aggressive, or extremely negative. In other words, they can be difficult. *Dealing With Difficult People* (Harvard Business School Press, 2004) and *Handling Difficult People* (Bloch, 2005) describe how to handle difficult people in the workplace, but as is common with many books on this topic, the suggestions in these popular works are not behavioral. As a supervisor with training in behavior analysis, you should find your job much easier for you to do than it would be for someone with an MBA or a job in human resources who has no clue about how behavior works. The business literature sounds somewhat behavioral with the general recommendation to "diagnose and prescribe" (Harvard Business School Press, 2004, p. 48) but falls far short of what you can do as a behavior analyst. You know, for example, that employees are much more likely to perform up to standard if they know what is expected of them and if they are properly trained for the job. You also know that providing regular feedback on performance is essential rather than merely a method of holding employees accountable (see more on this in Chapter 16, "Performance Management").

As an example, let's consider one of your supervisees whose work is not satisfactory. Behaviorally speaking, this is an employee who avoids or resists work assignments, often by making excuses, conveniently forgetting, arguing, working inefficiently, and procrastinating on most assignments. Rather than overtly saying to you, "I don't want to do this," this employee's actions speak for him, and thus he avoids an initial confrontation. When challenged, he (and others like him) will usually deny his behavior has this purpose and may turn the issue around to make the supervisor (in this case you) look guilty of being demanding or controlling. As a behavior analyst, you would not consider this a personality disorder but first want to know what the discriminative stimuli, S^Ds, are for the behavior. You would also want to examine the establishing operations connected with this type of behavior. Additional considerations include proper training for the task, the type of reinforcer available for completing the task, and the schedule of reinforcement. It is fairly obvious that a person you supervise who engages in this passive-aggressive kind of behavior has likely been doing this for some time and has probably been reinforced many times for behaving like this. You could ask yourself, "If I want to produce some appropriate work behavior in this person, where can I start?" This will most likely lead you to a shaping and fading program with a fairly dense schedule of reinforcement.

If there is a good thing about a problem with a direct report, it is that you might be in a position to control the reinforcers and consequences for this person. But sometimes, as a behavior analysis consultant, you'll be dealing with difficult people you don't supervise, such as your colleagues and peers.

DEALING WITH COLLEAGUES AND PEERS

The relationship you have with your peers and colleagues can present challenges if any of these people are difficult. Your primary consideration should be to protect yourself from harm by association with people who are negative, argumentative, burned out, overly dramatic, or

"Your primary consideration should be to protect yourself from harm by association with people who are negative, argumentative, burned out, overly dramatic, or two-faced."

two-faced. And, of course, you will need to design your professional life in such a way that you do not depend on people like this. These are people who can set you back in your professional development. We are working from an assumption here that your association with these peers is voluntary on your part. Everything changes if you are thrown together by your supervisor or boss and expected to work together as a team.

Using good behavioral skills such as reinforcing the person for good ideas and contributions may be helpful. It is also a good idea to keep your boss in the loop and objectively report on meetings that you have with the difficult person; for example, "I'll be glad to write up the minutes of our meeting and send them to our unit director." You should establish at the beginning of large projects which person is responsible for which tasks and document this in writing. Regular short meetings for updating on progress, with the difficult person having the chance to present her concerns, are also a good idea. Having witnesses to your interactions may be helpful if the colleague or peer is really a problem. If it is clear you have run into a difficult person who appears to be trying to do you harm, know that this kind of difficult person can be deadly if not dealt with effectively. You may have to involve your supervisors so that things don't get out of hand.

DEALING WITH UPPER MANAGEMENT: PRESIDENTS, VICE PRESIDENTS, AND CEOs

If the difficult person in your job is in upper management, all of the behavioral skills that we have discussed before apply. In addition to using these skills, you may want to set up a meeting with the person and ask if there is anything you can do to make this working relationship better. Explain your position in a calm, professional manner. If the person is a manager, director, section chief, or vice president and there are other people in the agency with the same job title, can one of them advocate for you or provide back up in meetings? Be sure to ask for feedback from others to make sure you are, in fact, on the right track.

If the difficult person is at the very top of the food chain, such as the president of the company, CEO, or agency owner, try everything that has been suggested earlier. Also ask for a meeting, and make every attempt to get things on the right track. If appropriate, your supervisor may need to be the one who meets with upper management on your behalf.

Sadly, sometimes it is the case that a good behavioral consultant and an agency are just not a good match. If the CEO or owner doesn't like you or your work no matter what you do, and your supervisor can't help, you may need to take a hint from the Kenny Rogers song "The Gambler": "You got to know when to hold 'em, know when to fold 'em,/Know when to walk away, and know when to run." We hope you'll know when to ask for help and will be able to fix any problems you encounter in the work setting. If the problems with difficult people in your work setting can't be resolved, remember that as a behavior analysis consultant, your skills are in demand. You deserve to have a happy life and work where you and your skills are appreciated.

ARE YOU THE DIFFICULT PERSON?

Before we leave this chapter, it is important for us to ask that you do a self-check to determine if you might actually be a difficult person from time to time. If you encounter others who are consistently resistant to your ideas, who argue with you about your strategies, or who appear to try to diminish your impact on a work group, it is time to determine if

> "There is some chance that in your zeal to help others you move too quickly, do not listen to objections, or provide too little support and reinforcement."

perhaps you are eliciting these reactions by your own demeanor. The best way to find out is to seek out a trusted colleague who will give it to you straight. Be prepared for some uncomfortable feedback, and do not engage in resistance or argumentation. There is some chance that, in your zeal to help others, you move too quickly, do not listen to objections, or provide too little support and reinforcement. Remind yourself that although implementing behavioral programs is easy for you, it can be very difficult for others. You might not have any qualms or concerns about using extinction or making reinforcers contingent, but other people might, and they need to be very comfortable with such methods before they will fully commit. One frequent reaction from nonbehavior analysts is that the consequences of implementing extinction or even contingent reinforcement is that they feel sorry for the individual and want to just give in. A common reaction from caregivers to a shaping program being implemented with a tantrumming toddler is "Wouldn't it just be easier to pick him up when he cries?" This is not surprising, but of course it will totally undermine your attempt to shape the toddler into a pleasant, cooperative child who listens to his parents.

If you are encountering resistance, argumentation, sabotage, or stonewalling, consider the possibility that this reaction to your approach to dealing with colleagues, volunteers, or clients is basically *your* fault; you need to do our self-check to determine if there is some other way of dealing with people.* You might be moving too fast or not be taking questions or concerns seriously, or you might assume that people know more than they do about how to implement your programs. With colleagues, you may not be giving enough reinforcement to maintain their contact with you. In your time with them, are you trying too hard to be the center of attention? If you focus on others and *you* provide the reinforcement, you may discover that they are much more willing to listen to you later. And if you are working with colleagues on a team project, do you always have to have everything your way? Perhaps being more of a compromiser would lessen the tension and eliminate some of the difficult behavior that you encounter each week.

---

* See the appendix at the end of this book.

DIFFICULT PEOPLE AREN'T EASY

It would be incredibly naive to assume you can simply start reinforcing selected behaviors of all difficult people and turn them around in a hurry. Some of these individuals have been honing their maladaptive behaviors for years. Some difficult people will lie about their work, you, and everyone around them. They prey on innocent staff (often at a lower rank than they are) and use them as pawns who can be easily manipulated. They take the ideas of others and present them as their own. They smile in your face in meetings and then run around the building and try to undermine and undo your projects. There is always someone they have targeted to "get rid of." The target is often someone with excellent skills whom the difficult person finds threatening. The target could be you.

YOU CAN'T WIN 'EM ALL

We knew a behavior analysis consultant who worked in a corporate setting. She was incredibly competent, had great social skills, and was liked by everyone. Everyone, that is, except one vice president, who could not stand her. She was friendly and upbeat when she interacted with him. She tried to use her best behavioral skills and reinforce him for his good ideas in meetings. We had seen this person in action, and her words of reinforcement and appreciation always came across as sincere. After a period of time, the consultant figured out that the vice president (a former military officer) interpreted friendly and cheerful behavior as "lightweight," particularly when a happy demeanor was exhibited by a woman. After some consideration, the consultant toned down her social skills when she was in the presence of this vice president. All small talk stopped. She never mentioned the weather or asked him how he was doing or how his family was, and she talked only about work tasks. She made an effort to actually smile less in his presence. This resulted in some improvement in the interactions, but there was another issue. The vice president wanted someone, a close friend of his wife, in the consultant's job. We asked the consultant how she had managed to survive so long. She responded,

> He may ultimately get me. But once I realized he will never like me, I needed a new plan. I make sure my work is excellent, even if I have to take it home and put more time in it at night. I copy my boss on everything, so if he claims I am not producing, my boss knows what is going on. I don't stop there. I send monthly reports to the president and other key staff. So if the vice president goes into a meeting and says I am not doing anything, he looks like an uninformed moron because everyone else at the table has seen my productivity for the past month. I also mentioned to a few key people that he was out to get me. I did this only once and then dropped it. It would not be professional to run around and act obsessed about this, but I did let people know he had his own agenda. If I get a written compliment from a customer or vendor, I pass it up the chain so people above me know I am an asset to the company. I volunteer for work that other people don't have the skills to do. I make sure my work is well documented all the way from the big projects to the number of e-mail and phone calls I handle. I am extremely respectful to this vice president in meetings, even though I am tempted to roll my eyes and mouth the words "blah, blah, blah" when he is talking. Basically, every day I work very hard to make myself a professional that my company wants to keep.

SUMMARY

Most of the time, dealing with difficult people is really no different from dealing with anyone else: Their behavior is under some kind of stimulus control, the establishing operations are there to be dis-

> "Most of the time, dealing with difficult people is really no different from dealing with anyone else."

covered, they come to the table with a history of reinforcement (probably not knowable) that can be assumed from the dynamic properties of their behavior, and, most important, their difficult behavior is modifiable through shaping. In situations such as the case just presented, the 25 essential skills presented in this book, along with sound behavioral skills, will help a consultant develop a strategy for dealing with difficult people.

Rather than getting emotional and labeling the difficult person as a "jerk" or other unprintable terms, the determined behavior analyst consultant can use the 25 essential skills along with techniques derived from 40 years of applied research to make some amazing changes in behavior.

FOR FURTHER READING

Bloch, J. P. (2005). *Handling difficult people*. Avon, MA: Adams Media.
Carnegie, D. (1981). *How to win friends and influence people*. New York: Simon & Schuster.
Harvard Business School Press. (2004). *Dealing with difficult people*. Boston: Author.
Hunsaker, P. L., & Alessandra, A. J. (1980). *The art of managing people*. New York: Simon & Schuster.

13

Think Function

For many people, one of the most frustrating aspects of life is not being able to understand other people's behavior.

Richard Carlson

I don't bring my work home. My time is my time. I leave work issues at the office. Some people separate their work from the rest of their life. This particularly makes sense for corporate lawyers, rocket scientists, and brain surgeons, for whom bringing work home would actually be considered aberrant behavior. Other than talking about a case with a trusted spouse, let's all hope the brain surgeon doesn't bring work home.

Other professionals could bring their work home, and it would not be considered the least bit unusual. Carpenters, plumbers, and nurses could easily use their skills at home to make the life of their family members better. Hanging new cabinets in the kitchen, installing updated modern bathroom faucets, and bandaging a knee injury all make sense. What about committed professional behavior analysts? Does it make sense that they would bring home their knowledge and skills about behavior? Should behavior analysts apply what they learned in graduate school when they are out and about in community settings? And what about with coworkers in the office? It certainly seems that the nature of our technology of behavior is applicable in all of these settings and that behavior analysts should certainly apply what they know about human behavior as opportunities arise. This doesn't mean the behavior analyst will always spring into action and begin shaping on behaviors. Sometimes, getting involved in the business of other people is simply not appropriate. The trained behavior analyst, however, should be able to observe a person's behavior in the natural setting and, to a reasonable extent, determine causal variables that everyone else seems to miss. Although most of the shoppers in a store look at a screaming child and think "spoiled brat!" the behavior analyst witnesses the same behavior and understands the mother has been intermittently reinforcing the child for tantrums in the grocery store: "Here's some candy. Please stop screaming!"

TROUBLE IN THE MALL

Amy and Amanda, twin sisters who are now all grown up, live in the same city. Even though they both have very busy professional lives, they try to get together at least once a month. Amy, an office manager, has a 2-1/2-year-old toddler, Jayden. Jayden is quite a handful. He began walking when he was 18 months old, and he has been babbling and talking for several weeks. Amanda, a Board Certified Behavior Analyst®, who is 3 months pregnant, has been thinking she'll be able to learn a few things about living with a young child from her sister. To Amanda's surprise, what she's learning is that Amy really doesn't have a clue about human behavior. The sisters are spending an afternoon with Jayden at the mall. They are cruising down the wide promenade of the mall with Jayden in the stroller, and the air is split by a high-pitched scream; Jayden is pointing to a kiosk where ice cream is prominently displayed, and the vendor is offering free samples. Amy veers to the left sharply and, nearly crashing into the young vendor, says, "OK, OK. We'll get the ice cream. Stop screaming, Jayden." As Amy tries to pull back into the stream of pedestrian traffic in the mall, Jayden begins to tantrum. As Jayden screams at the top of his lungs, turns red-faced, and bucks in his stroller, Amy says, "He wants more ice cream. He's like this all the time. It's the 'terrible twos'." Amanda is shocked that Amy has so little understanding of what she has just done and has no problem reconstructing what led up to this discomforting situation.

Looking at the situation from the viewpoint of a behavior analyst, Amanda thinks to herself,

> Here's what happened. They probably come here fairly regularly, because the mall is close to Amy's house. The first time Amy saw the vendor giving out free samples of "space-age ice cream," she thought it would be a cool thing for Jayden to experience. Amy got a sample for Jayden, they had a great time talking about "space-man ice cream," and off they went. Now, when Amy is in a hurry and doesn't want to stop, Jayden protests. To stop the screaming, Amy has been giving Jayden what he wants, just as she did today. The rest is history.

Amanda applied the "think function" motto when she was off duty. If she wasn't a behavior analyst, she would probably agree with Amy that Jayden was just going through a developmental stage commonly called the "terrible twos." The alternative explanation that Amy is shaping up some terrible behavior is not such a pleasant thought to have about your sister. And parents certainly don't like to think this about themselves. But that's the way it goes sometimes. Recalling that our code of ethics recommends against working with relatives, Amanda doesn't say anything to Amy, but if asked, she will explain that the "terrible twos" are a myth.

Countless well-controlled studies of child–parent interactions point in the direction of badly designed contingencies of reinforcement (not planned, by the way) by a parent who has no clue how behavior really works. Amy has an excuse; she's *not* a behavior analyst, she's an office manager, and she doesn't know any better. Our culture doesn't even come close to educating new parents about child behavior. The larger cultural contingencies of two busy parents and preschool-age children who are put under the stress of multiple trips back and forth to the day care center while sitting in the back seat of a Volvo by themselves do not help matters any. If parents seek behavioral assistance, we can help them understand that vague instructions,

> "Countless well-controlled studies of child–parent interactions point in the direction of badly designed contingencies of reinforcement ... by a parent who has no clue how behavior really works."

threats, and inconsistent, intermittent reinforcement for inappropriate behaviors are all likely to produce challenging emotional responses that will be quite demanding. A tantrumming child (who has been shaped to engage in this behavior) will usually come into contact with some negative quality-of-life issues in fairly short order. Yelling. Spankings. Restrictions. Before long, these consequences will start to add up to produce a child with an attitude toward his parents and maybe adults in general. It doesn't take psychotherapy or family counseling to get to the root of this problem, and a behavior analyst who is able to apply the think function will usually be able to find a behavioral explanation for such situations.

All behavior has a function. The function is right there in the environment and can be seen by a well-trained behavior analyst just as clearly as a northern spotted owl can be detected in an old-growth forest by an ornithologist with a good pair of binoculars. Just as biologists are trained to find rare species, behavior analysts are trained to see and understand contingencies of reinforcement. When we apply our professional training to the behavior occurring around us, we can spot the antecedents, imagine the establishing operations, and understand the consequences. What we usually don't know are the histories of reinforcement or the schedules; this takes a little more observing. But a behavior analyst who can apply the think function has a tremendous advantage over anyone from any other profession; it is easy to remain calm and objective if you understand how behavior works. Even if someone is directing some venom at you, it's not so bad. "Hmm," you say to yourself. "This guy must be having a really bad

> "Being an effective behavioral consultant means carrying your behavioral gear with you everywhere, analyzing each situation you come into during the day, and being prepared to apply the most appropriate tools when needed."

day. He's not always friendly, but this yelling is unusual. I wonder what happened to him this morning." As a behavior analyst, you have a very powerful edge on the outraged person. You know that to respond with anything but a blank look will reinforce it, and you also know that just as soon as he comes up for air, you can jump in and change the topic to something a little more pleasant. Being an effective behavioral consultant means carrying your behavioral gear with you everywhere, analyzing each situation you come into during the day, and being prepared to apply the most appropriate tools when needed.

THINK FUNCTION AT WORK AND AT HOME

So, what happens when the same behavior analyst goes to the office, work, school, or home to her own environment? Can she still apply the think function to effectively deal with her negative colleagues, nosy friends, bossy mother, hard-to-please clients, or thickheaded, obnoxious boyfriend? Absolutely.

Carol works for a behavioral consulting firm that provides services to exceptional student education (ESE) schools. At the end of the day, she stops by the office to write up reports, post her billing, and check e-mail. Janice is another consultant who often shows up about the same time. Soon after arriving, Janice begins to complain about her low hourly rate, the dreadful schools to which she's assigned, how frustrating the work is, and more. The first few times this happened, Carol was sympathetic and tried to console her colleague. "Well, these are hard times, and none of us are getting raises," she'd say. Or, "Yes, that is a tough school, but there are some nice teachers there." After complaining to her husband about Janice, "She's starting to get on my nerves. I'm tired of listening to her, and I wish they would reassign her," Carol at that point realized this was a *behavioral* issue: "Somebody in the office probably starting reinforcing it with and here I am maintaining it with sympathy." That was a turning point for Carol. Deciding that Janice's whining was attention driven, Carol began to put all whining on extinction while reinforcing Janice for other topics of conversation. The whining soon stopped when Carol was present.

The solution to the problem between Carol and Janice seems so obvious. How could a behavior analyst not instantly recognize inappropriate verbal behavior and know that it should be ignored? The answer seems to be that we behavior analysts are mere mortals, and as such some of us forget from time to time that we can and should apply our behavioral knowledge in our everyday life. There is some irony in the situation where Carol complained to her husband about Janice's tendency to complain. Yet, many of us get carried away and allow ourselves to be indignant rather than effective, or we swear rather than shape on a behavior.

THINK FUNCTION IN THE WORKPLACE: READING SUBTLE SIGNS

The principal is not smiling. David is the consulting behavior analyst in a school. He reports directly to the principal. Normally, the principal is cheerful and happy to see David, but something has changed. Fortunately, David has been observant and noticed the change. The principal won't make eye contact, she has been short in her responses to everyone, and she is making no positive comments like she normally would.

David needs to start thinking about all of the possible functional variables:

- It's probably not something he did, because she's not smiling at *anyone.*
- Could there be a looming aversive stimulus in the principal's life that David is not aware of?
- Could there be a motivational variable that has changed for her?
- Is it possibly a biological or medical factor (is she sick)? Has she received some bad news from her doctor about a mammogram?
- Has an important reinforcer been removed? Did her husband leave her? Did she hear that she is being transferred to another school?

David understands that the principal's behavior is a function of something. It's not just random, and he needs to be concerned and cautious. If something dreadful has happened in her life, he needs to show caring and compassion. This is not a good time to bring up that sticky issue about the first-grade teacher who has been yelling at her kids.

CASE OF THE CHATTY CUBICLE MATE

In a corporate setting, Sharon shared a cubicle with Pam, another behavior analyst. Day after day, Pam would come in, sit down, turn on her computer, and begin to work. Sharon said,

> Pam has great work habits, but she is strange. She is nearly mute, never exchanging pleasantries, asking how my weekend was, or saying good-bye when she leaves. Then, all of a sudden, today she started acting like I was her best friend. She kept trying to start a conversation, and get this! She wanted to know if I would like to go to lunch with her. It's so weird, I'm not sure I want to go. What do you think she's up to?

Possibilities include the following:

- Something *good* has happened to Pam that she wants to share.
- Something *bad* has happened to Pam that she wants to talk about.
- Pam has discovered a reinforcer that Sharon holds for her.

Can you think of any other functional variables that might be affecting Pam's behavior? Can you give specific examples of these variables (e.g., as an example of something good happening to her, Pam might have been promoted)?

HOSTILE INFORMATION TECHNOLOGY DIRECTOR (THE COMPANY COMPUTER GURU)

Bill is a behavior analyst who works as a performance management specialist for a large national insurance company. The company has a director of computer services who basically has the final say on which departments in the company will receive computer support services. If you need a new software application, Mr. Guru is in charge. If you are having problems with an application, Mr. Guru decides if tech support will help you. Bill thought he was staying under the radar when it came to Mr. Guru, but at a meeting with company management, Mr. Guru criticized Bill's work and attempted to have one of Bill's projects cancelled. "It's a waste of company time and money," said Mr. Guru. Bill left the meeting feeling somewhat dazed and wondering what was behind Mr. Guru's behavior.

Possibilities include the following:

- Mr. Guru has it in for Bill.
- Mr. Guru has it in for Bill's boss.
- Mr. Guru is trying to intimidate Bill for some reason.
- Mr. Guru is trying to impress someone at the table in the meeting.
- Bill has been unrealistic about his project and skills.

Can you think of any other functional variables that might be affecting Mr. Guru's behavior? Can you give specific examples of these variables? In the work setting, when someone says something like "Mr. Guru has it in for Bill," how do you translate that into behavioral language? What could Mr. Guru's motivation be for "getting Bill"?

DISTANT PARENTS

As a behavior analyst who works in the home, Betsy enjoyed working with Justin's parents. They were upbeat, optimistic, and ready to try anything Betsy asked to help Justin. But lately things have changed, and Justin's parents have been rather somber. They show little enthusiasm for any of Betsy's suggestions. Betsy has noticed they seem to avoid eye contact with her and each other, and although they have been polite, their politeness seems forced.

Betsy has been thinking of some of the things that could cause a change in the behavior of these parents.

Possibilities include the following:

- The parents are no longer happy with Betsy.
- The parents are no longer happy with each other.
- The parents have decided they are not up to dealing with Justin at home.
- The parents are having financial difficulties.
- There could be medical issues.

Can you think of any other functional variables that might be affecting the behavior of Justin's parents?

SUMMARY

As a behavioral consultant, you need to be aware of your social environment at all times. Applying the think function will help you read the cues to interact effectively with people, even when you are off duty; the behavior of people in the world around you is lawful. There are controlling variables for all behaviors, and as a professional who is committed to the field of behavior analysis, you need to try to determine what the variables are and make the necessary adjustments or corrections so you can respond appropriately and professionally.

FOR FURTHER READING

Bailey, J. S., & Burch, M. R. (2006). *How to think like a behavior analyst: Understanding the science that can change your life.* Mahwah, NJ: Lawrence Erlbaum Associates.

14

Use Shaping Effectively
___

Shaping, shaping, always shaping.

**The tagline of a top-selling product of the
Florida Association of Behavior Analysis**

*Out of frustration, a teacher calls on a third-grade student
who is frantically waving her hand and quietly but intensely
saying, "Me, me, me!"*

*On a family trip to New York, a dad, after saying, "OK, I'll
take you to the American Girl doll store, but we're not buying anything today," gets out his platinum credit card and
gives in to his begging 9-year-old daughter.*

*An arborist supervisor, busy with an important cell phone call,
notices that her tree-trimming workers have failed (again!)
to put on their hard hats, but she hurries to her truck, saying nothing.*

The laws of behavior operate whether or not a person is a behavior analyst. In each of these three cases, someone important
has engaged in the shaping of inappropriate behavior. The behavioral offenders didn't realize it at the time, and they probably
wouldn't admit it if it were pointed out to them. But just like these
seemingly innocuous cases, there are interactions occurring millions of times each day in our culture that add up to a chronic
behavioral headache for everyone.

The third-grade teacher is now confronted with a student who insists every day on being called on, urgently waving her arm and hissing, "Me, me, it's my turn. I know the answer. Puh-leeeze!" Dad has a preteen on his hands who is constantly begging for something, and the owner of Miller's Tree Service has a workers' compensation claim to deal with from an employee who sustained a head injury when a falling limb knocked him silly on Friday afternoon.

In each case, the principles of behavior were operating to ever so slightly increase the likelihood of an inappropriate behavior. With this increasing probability of occurrence goes an even greater chance of reinforcement, in an ever-increasing circle of settings.

> "The ubiquitous power of reinforcement, applied in small dribbles and drabs throughout the day, produces the complex scene of human behavior that unfolds before us as behavior analysts."

The ubiquitous power of reinforcement, applied in small dribbles and drabs throughout the day, produces the complex scene of human behavior that unfolds before us as behavior analysts. Our behavior analytic world is made up of complex, difficult behaviors that require functional analyses to sort out, written behavior plans to change, and formal training protocols, and it often requires patience, tons of patience, to change behaviors. In this sea of accidental reinforcement, the behavior analyst is tossed about by the waves of inappropriate behavior that will swamp his kayak if he's not careful. What is a well-trained, conscientious, law-abiding behavior analyst to do?

ETHICS OF SHAPING

In the hands of a well-trained behavior analyst who has excellent professional skills, shaping can be a joy to behold. Consider the behavior analyst who is in a group of people that includes a very difficult person. The behavior analyst engages in conversation easily, smiling and nodding her head frequently in conversation. Someone watching the interactions will notice that she might occasionally pause, show

> "Our behavior analyst is calm and in charge. She has no reason to get mad or argue with anyone because she knows the behavior she is seeing is the product of the person's past history, probably going back to childhood."

a neutral expression on her face, and then become engaged again. Amazingly, the stable, popular behavior analyst is able to make instant calculations as to the likelihood that the behavior she is witnessing is going to cause some trouble down the line. She will apply the smiles, agreements, and head nods at precisely the right time to shape on the behavior. Our behavior analyst is calm and in charge. She has no reason to get mad or argue with anyone because she knows the behavior she is seeing is the product of the person's past history, probably going back to childhood.

The complainer, the know-it-all, and the drama queen (Bloch, 2005) all present difficulties for everyone around them except the well-grounded behavior analyst, who understands how behavior works. There is a Zen essence to this deep knowledge that gives the behavior analyst not only the calmness that comes with understanding but also the permission to act as needed. Shaping can be done accidentally by the untrained and deliberately by the well trained. Although using behavior analysis in their daily lives is not a requirement, it certainly should not be surprising to discover that individually, behavior analysts can use their knowledge to create some degree of appropriate behavior in their realm of influence. Applying what you know about the basic principles of behavior to the people around you is in your, and their, best interest.

DEVELOP THE SHAPING HABIT

Habits are behaviors that come naturally. They are often described as "unconscious" behaviors, meaning that they are not reported verbally.* Behavioral consultants interact with dozens to hundreds of people each week. The opportunities for shaping are enormous, but, as with most acquisition tasks, it is probably best to start small. In the beginning you will need to select one or two people and then one or two behaviors for each person you have selected. With shaping, you are not trying to produce a totally new behavior but rather trying to increase the likelihood or probability that some response will occur more often or at the right time.

As an example, assume that you are the consulting behavior analyst who has been hired by an administrator to work with his staff.† There are several problems you'll need to address with the staff, including skill deficits related to training clients, off-task behavior, and a lack of motivation. Your first task is to convince the administrator that the intensive behavioral makeover of his staff will take some time and that it cannot be accomplished by a short round of stand-up training. You will need to shape *agreement behaviors.* So what are they? Ultimately you want to hear the administrator say, "That sounds good. I agree. Let's go forward with your plan." But, using shaping, you can begin with an approximation of your final goal by saying to the administrator, "I thought I could call a meeting for Friday afternoon, and maybe you could just talk to the employees about their *attitude* toward the clients."

Clearly, there is a big distance between having the employees change their attitudes and changing their actual behavior. So if you start talking to the administrator about a short meeting to talk about attitudes, how do you get from A to B? The smart thing to do is identify something with which the administrator will agree, some comment that will prompt head nodding: "I know

---

* "Unconscious" just means untacted (unlabeled) behavior.
† We assume that you have already established yourself as a reinforcer, as described in Chapter 1.

you've had this problem for quite a while. You must be getting tired of dealing with this." (You're showing empathy for his situation and increasing the likelihood of his nodding in agreement.) He nods, sighs a big sigh, and says, "Yes, that's right. It does get old." You sigh too (matching and mirroring), say, "I know how that is," and then relate a story of dealing with a neighbor or someone in another work setting, then come back to his situation: "My experience has been that changing a long-standing behavior can't be done overnight." Look for a reaction or any sign of agreement. If it occurs, reinforce it with a head nod and comforting words: "It really takes patience to be an administrator. I know it does." And so it goes. Remember that you should not just blurt out your entire proposal at the beginning of the meeting like a nervous amateur. You present it carefully and slowly, bit by bit, reinforcing agreements, head nods, smiles, or confirming comments: "Well, I guess you're right. Now that I think of it, our other consultants have tried pep talks and motivational speeches, but the effect is really short-lived." You agree heartily and point out that there really is no research that shows that these strategies have ever worked. You say the pep-talk approach is more like a photo op to you: "It's all hat and no horse," you say. (This goes over big in west Texas. Note to self: Learn local idioms.) He laughs. You laugh. Why are you reinforcing laughing? Because it is paired with head nodding, smiling, and agreeing with you. Remember why you are here: You need this administrator to agree that your proposal for four hands-on training sessions, spaced 1 week apart, with a maintenance and generalization test, is absolutely the best thing he can do under the circumstances.

When you are beginning your consulting career, each of these encounters will seem painful. You'll have to be so on your toes, watching so intently for each little motor movement indicating the beginning of a smile or a slight glint in the eyes, that it can give you a migraine. You need to prepare carefully for such meetings, analyzing your pitch, breaking it down into small steps, and memorizing it until you can do it without notes so that it comes out like the totally natural conversations you have every day. On top of that, you must focus on every little movement of the eyes, small hand gestures, and shifting of positions, looking for any sign of agreement or at least a break in resistance. The administrator's moving from "Well, I don't know" to "Well, I guess" deserves a *big* reinforcer from you.

After a few of these meetings, it should become second nature to use shaping as part of every presentation and in every interaction with your clients. There will be times in meetings when someone will present troublesome behavior. This is when your blank face comes in; you

> "As you are developing your skills, shaping on others can be exhausting, but with time, it will become automatic, a habit. A good habit."

need to be able to turn it on and turn it off again in a split second. Not a frown, no head shaking, just a blank face, for just a second. Someone made a comment that wasn't helpful. You respond with no reaction, wait a few seconds, and then go back to your presentation, smiling and nodding your head. As you are developing your skills, shaping on others can be exhausting, but with time, it will become automatic, a habit. A good habit.

EVERYDAY BEHAVIOR SHAPING

In addition to shaping clients, you will probably find yourself faced with numerous opportunities every day to use shaping with your supervisor, colleagues, and family members (Sutherland, 2008). Is it ethical to begin using what you learned in school with the people in your daily life? With just a few qualifications, the answer is *yes*. The effect of reinforcement, in the right hands, can be truly powerful. If you hold a reinforcer for a person, know how to use conditioned reinforcers, and are good at your timing, you can deliberately shape some amazing behaviors. But you need to think through a few issues first. You cannot use this knowledge, this deep understanding of human behavior, to promote your own personal agenda. It's not fair, ethical, or appropriate to differentially reinforce your friends for buying you beer, washing your car, or walking your dog. It's just not right. If you are going to use shaping with the people around you, you must always be ethical about it by selecting behaviors that are in *their* best interests.

For example, suppose you strike up a friendship with someone and discover you have a lot in common, you get along well, and you enjoy spending time together. But this person has an annoying habit of interrupting. It's not malicious, it's not a deal breaker, but you wouldn't want to get caught in an elevator over a weekend with this person, because the constant interrupting would drive you crazy. What should you do? You don't have to bring this annoying habit to the person's attention. Most likely it wouldn't do any good, and in fact, calling attention to the problem could set you back. Interrupting while others are talking is not a cognitive problem, and talking about it will just make it worse. Shaping is the answer.

Making shaping look natural so that no one accuses you of being manipulative is a requirement, not an option, for behavior analysts. You will lose friends in a hurry if they think you are attempting to trick them into doing something. Your goals must be honorable and justifiable, not self-serving or demeaning.

> "Making shaping look natural so that no one accuses you of being manipulative is a requirement, not an option, for behavior analysts."

Sutherland (2008), in her excellent book *What Shamu Taught Me About Life, Love, and Marriage*, described using social reinforcers for all of the people around her, from the impatient person in line behind her to her mother who needed to quit smoking and her husband who constantly lost his keys. Sutherland had no qualms about her strategy and was willing to go public with her plans. The media response to her book was overwhelmingly positive. It seemed that most people were enthralled with the idea that one could use the same behavioral procedures used to train animals to improve relationships. In one particularly telling example, Sutherland described how, prior to her use of shaping at home, she enabled her husband's frequently losing his keys by sympathizing with his plight and joining in the frantic search. When she realized what she was doing, let him solve his own problem, and simply reinforced him when he found the keys, the losing became less frequent, and the frenzied searches became calmer and more systematic.

You will be a far more effective behavioral consultant if you can develop the habit of looking for behaviors to reinforce wherever you go. At work you may have colleagues with annoying habits such as borrowing your stapler and not returning it, coming to meetings late, or telling off-color jokes. Most people don't seem to know what to do in certain situations, so they engage in thoughtless responses that just perpetuate the problem. Saying, "That's OK" is not a good consequence for someone who has done or said something inconsiderate. Rather, a systematic, consistent plan to reinforce closer approximations to considerate, responsible behavior is in order. Dealing with inappropriate jokes calls for giving the blank look, holding it for a couple of painful, agonizing seconds, then changing the subject to prompt a more appropriate behavior.

Mediators (e.g., teachers, parents who will be carrying out behavioral treatment plans with a child) are particularly sensitive and responsive to well-calculated shaping. They are trying hard and want to please. In their new role as change-agents, they may be desperate to know if they are doing what is expected. They are a long way from actually changing a behavior and coming into contact with the natural consequences of their new strategy. The best form of shaping is that which is delivered concurrently with their attempts. Let's look at the situation where the mediator is the mother of an autistic child. You've been trying to teach her to conduct language sessions with the child. You'll want to position yourself so you can observe the interaction between the mother and the child. So you can do the shaping that is needed, sit or stand so the mother can see your face and easily make eye contact. Mom will be able to see your head nod as you encourage training attempts. When she glances your way, a big smile and silently mouthed "Terrific!" or "Great!" will have the desired effect. At the end of the session, you can follow up with immediate, descriptive feedback to show the mother how pleased you are with her progress. You'll be not only increasing the rate of correct responding but also building the confidence and self-assurance that will be needed in the upcoming days when Mom hits a rough patch and you are not there.

SUMMARY

"Shaping, shaping, always shaping." This mantra repeated slowly and occasionally throughout the day with your voice pitched an octave lower than normal serves as a reminder that you, as the behavior analyst consultant, *can* influence the behavior of people around you. A client (or family member) who is being hardheaded and won't listen to reason, your nosy cubicle mate with the too-personal questions, or the mediator who can't quite get the hang of her reinforcer timing will all improve with your systematic use of shaping. As a behavior analyst, you have all the skills needed to change behaviors. You know how to identify reinforcers, you control the reinforcers that *you* hold, and you have spent countless hours practicing your delivery so that it is perfectly timed for maximum effect. You can't lose as long as you are patient. As long as you remember "Shaping, shaping, always shaping," the laws of behavior are on your side.

FOR FURTHER READING

Bloch, J. P. (2005). *Handling difficult people*. Avon, MA: Adams Media.
Sutherland, A. (2008). *What Shamu taught me about life, love, and marriage*. New York: Random House.

15

Can You Show Me That?

The Key to Effective Consulting

Can You Show Me That?

Aubrey Daniels (2000)

Ethan did it again last night. I told him it was time for bed, and he absolutely ignored me. He kept clicking away on his Nintendo and mumbled, "Oh Mom, I'm almost done. Just one more level." You've helped me with his other behavior problems. What can I say to him that will make him realize that I'm serious?

As a consulting behavior analyst, you will witness this scenario nearly every day. Mom is desperate, because this has been going on for a long time now, and she is able to see the pattern. Ethan has a mind of his own, he won't listen, and he appears to be "limit testing." Ethan's mother needs help and will do whatever you say. It's tempting, it really is. Like a doctor prescribing pills, you can prescribe the treatment and make mom happy. She thinks you are *the* expert. You don't want to let her down, and all you need to do is tell her what to do. The problem is that you don't know. All you have is a frantic mother's description of her child's behavior and her version of what happened, but you don't have the facts. You weren't there, you don't *really* know, and in good conscience you can't simply take a guess at an answer.

"CAN YOU SHOW ME THAT?" FIVE WORDS THAT CAN IMPROVE THE BOTTOM LINE

Aubrey Daniels, the CEO and founder of Aubrey Daniels International and widely acknowledged father of performance management, confronts business and organizational consulting problems every day. CEOs of Fortune 500 companies hire Dr. Daniels to tell them how to motivate their employees, how to improve customer service, how to enhance employee awareness of safety issues, and more. Sitting in comfortable leather chairs in the boardroom, executives recount tales of employees who don't seem to care about the customers' needs. They describe situations where departments or divisions within the organization cost the company millions of dollars with their wasteful practices. These executives are important people with pressing problems, and they would like immediate answers. Preferably *easy* answers. Just like Ethan's mother in the scenario opening this chapter, these highly intelligent, powerful people are frustrated, and you can hear it in their voices: "Our new microchip assembly plant out in Chandler was supposed to be the answer to our production problems. It cost us $50 million and took 2 years to get up and running, and now it can't meet production goals. Don't they realize how important this is? We're in a global economy; Taiwan is winning. You're the expert on human performance. Tell me what to do."

Dr. Daniels not only has the credentials and experience to advise these captains of industry but also has nearly 30 years of research on his side. The *Journal of Organizational Behavior Management* has published countless studies showing how the basic principles of behavior can be applied in industrial settings. But Dr. Daniels is wary of easy answers; he has been bitten before and has learned from experience. It's not that the CEO requesting a quick fix is not serious (he is) or that he is confused (he isn't). But in describing what he sees as the problem, he's passing along hearsay to someone who has been in this situation more times than he can remember. The sincere and the desperate really don't understand how human behavior works. They tend to gloss over the details, they ignore facts that are critical, and they are blind to influential contingencies that operate just out of view of the untrained eye. Dr. Daniels knows just what to say at this critical juncture: "Can you show me that?" It can be a conversation stopper. It needs some elaboration of course, but in those five simple words he captures a key feature of effective behavioral consulting: We actually have to see what is going on to be able to do anything about it. Vague descriptions such as "They just don't seem to care" beget vague interventions such as motivational programs or, even worse, motivational *speakers*. Lack of awareness of safety issues normally leads to "awareness enhancing" videos, posters, and billboard-sized banners proclaiming, "Safety Is Job One" or "Bee Safe" (with grinning bumblebees wearing hard hats and goggles, flying around in the cartoon prominently displayed in a corner of the poster). "Can you show me that?" or "Can I see that?" What elegantly simple requests! So reasonable, so self-correcting, so spot-on behavioral. In its simplicity, the question "Can you show me that?" speaks volumes: "I don't trust your account. I need to see this for myself."

Experience has proved time and again that when it comes to human behavior, most people don't know what to look for. They miss the important variables, and they conveniently leave out key elements. Many people exaggerate what they have seen, or they go to the other extreme by minimizing something that is happening that is potentially dangerous. In the case at the beginning of this chapter, Ethan's

> "Experience has proved time and again that when it comes to human behavior, most people don't know what to look for. They miss the important variables, and they conveniently leave out key elements."

mom left out the important fact that she promised Ethan that he could play his computer game for 30 minutes if he completed all of his chores and homework. She left out the fact that she was trying to get him to go to bed early so she could watch *Desperate Housewives.* "*He* challenged me." "*He's* defiant." "*He* won't listen." You learned back in elementary school that there are two sides to every story. The classic 1950s Japanese movie *Rashômon,* in which the story of a horrible crime and its aftermath are told from different points of view, teaches the valuable lesson that there can be three or four sides to some stories.

SEEING FOR YOURSELF

The cold, hard truth for you as a behavioral consultant is that you don't know and can't understand a human performance problem until you've seen it for yourself. You know what to look for and have been trained to understand behavioral concepts. Concepts such as establishing operations take on special significance when you actually visit the microchip factory in Chandler and learn that the careful, slow-moving employees in their white hazmat-looking uniforms are paid *by the hour* and are stiffly penalized for any errors on the production line. You learn later that it isn't this way back in Cincinnati, where there are bonuses for high rates of *perfect* microchips. The CEO left out this little detail when he described the problem to you. He did this not out of malice or ignorance but because he didn't understand what difference this detail could make. This particular CEO believes he and all of his employees should stay focused on the "mission." "As long as they understand *the mission*, they should work hard for the company," he says. That's *his* mantra. Asking, "Can you show me that?" puts the behavior analyst consultant right where she needs to be so she can look at the environment where the action takes place. What *are* the antecedents for the desired behavior? The consultant can take a look for herself and see if the antecedents are likely to be effective discriminative stimuli (S^Ds). Is there any response cost associated with the desired behavior? She can ask, "Can you have him do that again? I'm not sure I understand exactly what happens when he does that last test." And she can ask realistic questions that didn't come to mind when she was sitting in a leather chair in a plush boardroom: "Tell me again about how they were trained and about that penalty for defects. Does that come right out of their paychecks?"

FROM THE BOARDROOM TO THE KITCHEN: TEACHING SUPERVISION TO MID-LEVEL ADMINISTRATORS

As part of a recent long-term consultation at a mental health facility where department heads were being trained to improve their effectiveness, another prime example comes to mind. The mid-level administrators who signed up for the leadership seminar were to take on a performance dilemma with their direct reports. In one case, it involved six hospital kitchen supervisors who were all long-term employees of the institution. The department head in charge of the kitchen supervisors was ready to give up: "I can't get them to supervise the line staff. They get into arguments with them, and they won't hold them accountable for anything. I might as well fire the whole worthless lot of them." The administration was routinely confronting the following issues:

- Line staff members were calling in sick when they just wanted a day off.
- The supervisors were chastising the line staff members on the phone for being lazy when they called in to say they would not be coming to work.
- Not a single supervisor had written up anyone for anything in the past 6 months, even though there had been a major meltdown in quality control, and morale was in the dumpster right along with the cafeteria garbage.

The behavior analysis consultant asked, "Can I see that?" and got a dumbfounded, blank look from the administrator for his well-timed, incisive question. She asked, "What do you mean?" "I mean I'd like to come over to the kitchen and see how you supervise your supervisors. And then I'd like to watch them do some supervision with the kitchen staff," he responded. "Like when?" "Like now!" And off they went to the sprawling, steamy, slippery-when-wet, odoriferous institutional kitchen where the staff cranked out 1,500 tasty, nutritious, medically approved meals each and every day of the year for 500 mental patients.

> "Off they went to the sprawling, steamy, slippery-when-wet, odoriferous institutional kitchen where the staff cranked out 1,500 tasty, nutritious, medically approved meals each and every day of the year for 500 mental patients."

As is commonly the problem, the description provided in the department head's overview of her operation was vague. She supervised them "all the time." She gave "immediate feedback." She was perfect. Direct observations in the noisy, hothouse-like kitchen area, however, revealed a different story. Supervision, such as it was, consisted of supervisors being occasionally called into the department head's office for a chewing out or pep talk based on the many complaints and rumors that were flying around the cavernous ovens, sleeping-bag-sized griddles, spotless stainless steel sinks, and whistling, steaming dishwashers. This form of supervision rolled downhill from the department head to the line staff. There were periodic futile attempts at behavior change with the most disgruntled staff.

SUPERVISING MEANS WATCHING PEOPLE WORK

The behavior analysis consultant continued his line of questioning: "Where do you stand when you supervise?" The response to this simple question was, "What?" He replied, "You know, if you want to supervise people, you have to watch them work and actually do their job. Where do you stand when you do this?" The blank look the administrator gave him told it all. She had never actually observed any of the staff as they did their job. It never occurred to anyone that supervision involved observing on a regular basis. After observing staff members, supervisors need to give people prompts, instruction, feedback, or, we hope, a positive reinforcer. We talked about this in the seminar. Everyone took notes, including the department head in question. There were head nods all around when we asked them, "So everyone agrees that this is the working model for supervision. Any questions?" Asking, "Can I see that?" and "Can you show me that?" opened up a whole new world of training. On-site training began immediately for the department head and her supervisors. They learned where to stand, what to look for, how to approach a kitchen worker and prompt the behavior that they wanted, and how to give feedback.

MODELING REINFORCEMENT

As part of the on-site walk-through, the behavioral consultant, standing shoulder to shoulder with one of the supervisors, asked, "Now what does she do there, the woman with the blue bandana on her head?" "That's Nadine. She's the inspector. She compares what's written on the dietary prescription for each patient and makes sure that each food item matches perfectly. We can't have diabetics getting pudding or our heart patients getting salty foods. It's not healthy. It's actually dangerous." "So here in the kitchen, she's an important person?" "Oh yes, we'd be in real trouble if we lost Nadine." We continued to watch as Nadine closely inspected each tray, occasionally removing a small piece of yellow cake with chocolate frosting and replacing it with a banana. "Do you ever check her work?" the consultant whispered. "Sure." "I'd like to see that. Do you mind?" The supervisor was happy to show how she checked Nadine's work. Taking a quick step behind Nadine's station where the trays went quickly into rolling carts, the supervisor looked at five trays. All five trays were perfect. The supervisor stepped back. "No errors," she said smiling at the consultant. "So now what?" he asked. "What do you mean?" "Are you going to do anything?" "No, she didn't make any mistakes. I've got no problems with her." The consultant said, "OK, watch me closely." He stepped over to Nadine and in a low voice said, "Looks to me like you are perfect, Nadine. All of your trays have been set up just right because you paid close attention and caught some mistakes. Thank you for being so diligent."

The consultant took a few steps back to the observation spot with the supervisor. "Any questions?" "No." "Did you see what I just did there?" "You told her she was doing a great job. So?" "Have you ever done that? I mean as part of your supervision?" You can imagine the rest of the conversation. The question "Can I see that?" led to a totally different understanding of the problem at hand. In the comfort of the seminar room, *supervision*, as a key element of leadership, was treated as a generic amorphous concept. It wasn't real. Half a dozen training sessions, including PowerPoint slide shows, the clear display of data, discussions, note taking, and Q & A sessions, revealed none of the rich details that were captured by direct observation in the actual setting. It is absolutely clear that the classroom training could probably never really affect what was happening in the kitchen, because the seminar leaders did not even know the specifics of how the operation worked. The request "Can I see that?" showed clearly what was happening on the ground and lead to very specific recommendations that were followed up in the next weeks and months with additional visits to the kitchen.

WHAT IF THEY CAN'T SHOW YOU?

If you can't directly observe the behavior on which you are asked to consult, what should you do? This is not a matter to be taken lightly. Advising people on human behavior problems that you cannot observe directly is akin to doctors prescribing medications without seeing patients. Doctors consider this unethical and will insist that the patient come to the office for a visit. Physicians understand that patients are notoriously unreliable in reporting symptoms. Patients often leave out key details that they didn't notice or consider significant, and of course they can't observe infections or an irregular heartbeat.

In addition to seeing the actual behaviors in action, behavior analysts also need direct knowledge of the environment in which a behavior occurs. Furthermore, trained behavioral consultants are more likely than an untrained person to gain important information from watching individual performances or interactions between managers and employees. Parents, teachers, and supervisors frequently overreport incidents just for dramatic effect. To get attention or access to resources, they may exaggerate the frequency or intensity of a behavior.

If you have been asked to provide consultation, and you have not been able to directly observe the problems, you might consider two options before you accept or turn down the assignment. First, ask if it is possible to get a video camera in the setting that is posi-

> "If you can't see the behavior directly, you won't be able to see the intervention directly either."

tioned in such a way that it will capture what is being described. This is certainly not as good as a live observation by you, but it does give you more information than the hearsay you are getting otherwise. What you are missing is all the ancillary information that is out of the view of the camera lens, such as antecedent events that precede the behavior, establishing operations that have occurred, and the fine grain elements of the behavior itself that can't be picked up at a distance. You will have to decide if you can operate ethically under these conditions and if you can give advice that will be followed carefully enough to establish integrity. Remember, if you can't see the behavior directly, you won't be able to see the intervention directly either.

The second fallback position is to rely on carefully collected data that you can graph and analyze. This works only if you have reliable observers using tested protocols. You will need multiple measures and repeated measures over time. Again, you will have to determine if you can operate in this ethical gray zone where you don't have direct access to the environment where your intervention is to be tested.

SUMMARY

To solve behavior problems in any type of setting, ranging from children at home to schools to corporations and institutions, behavior analysis consultants should directly observe problem behaviors. Supervisors and administrators should be taught that supervision requires direct observation, feedback, and plenty of positive reinforcement for a job well done. Finally, the use of the request "Can I see that?" will provide the consultant with an understanding of the *real* problem and the information needed to determine the best solution.

FOR FURTHER READING

Daniels, A. C. (2000). *Bringing out the best in people: How to apply the astonishing power of positive reinforcement.* New York: McGraw-Hill.

16

Performance Management

Performance Management is a systematic, data-oriented approach to managing people at work that relies on positive reinforcement as the major way to maximize performance.

Aubrey Daniels

The basic principles of behavior as we know them are no different when applied directly to an individual client than when they are applied to the entire staff of a group home or the workers in a factory. Target behaviors when you are dealing with workers are different in that they are not clinical problems, that is, you are dealing not with maladaptive, self-injurious, or aggressive behavior but rather with simple off-task, unproductive, wasteful, or unsafe behaviors. These behaviors, all considered normal for almost any work environment, are incredibly common, and they drive CEOs, managers, and supervisors absolutely crazy.

When these work-related problems occur in developmental-disabilities residential facilities, they can result in problems that can shut down an entire program. In some cases, serious problems with staff and a lack of good management can put an entire organization at risk of losing funding. If staff members are not doing their job and providing quality programming (and supervisors are not making this happen), the ultimate cost can be substantial sums of money, especially if the program leads to lawsuits resulting from neglectful or abusive behaviors. In business settings, these common behavior and management problems can cost huge amounts of money when added up across a fiscal year for hundreds of employees.

BEHAVIOR ANALYSIS IN BUSINESS, INDUSTRY, AND ORGANIZATIONAL SETTINGS

When behavior analysis principles are applied in business, industrial, or organizational settings, the term *performance management* (PM) is used to describe this area of application. PM is a specialty of the larger field of applied behavior analysis (see Daniels & Daniels, 2004). This specialty has its own journal (the *Journal of Organizational Behavior Management*), its own conference (the Florida Association of Behavior Analysis Organizational Behavior Management conference, known as the FABA/OBM conference),[*] and its own special interest group (the Organizational Behavior Management Network within the Association for Behavior Analysis International).

PM operates very much like the clinical and educational applications of applied behavior analysis. The process starts with a referral, usually from someone at mid-management level or higher in an organization, and may begin with an inquiry to a consulting firm. Another starting point might be when a behavior analyst who is working in a clinical setting notices that individual behavior programs are not being carried out. The behavior analyst suggests to the administrator that considering larger, systems-level interventions might be appropriate.[†] Our experience has shown that the initial referral to the behavior analyst consultant from upper-level management is often vague: "We have 'motivation' problems here that we need to address," "Our people are just not focused enough on safety," or "We need some major attitude adjustments in this

---

[*] This conference is held biannually in the odd years. The January 2009 conference was held in Cocoa Beach, Florida.

[†] Executive coaching is another PM application that is catching on. This involves working with upper-level management and teaching executives ways to be more effective in decision making and their interactions with various department heads. See Chapter 24 for more information.

company if our merger is going to be successful." There is a special art to identifying the actual problem. Aubrey Daniels is the individual credited with starting the field of PM in the mid-1960s. Daniels has referred to the identifying of an actual behavioral problem as *pinpointing*, that is,

> "Our experience has shown that the initial referral to the behavior analyst consultant from upper-level management is often vague."

converting a vague reference to a problem into observable, measurable behaviors. To identify target behaviors in PM settings, the consultant first must ask for examples of motivation problems so that specifics can be determined. For many managers or executives, motivation problems involve employees who come to work late, are off task, or are unresponsive to requests for information. For others, an employee motivation problem might mean employees don't complete tasks on time or they are sloppy in their work and waste expensive materials. The statement "My workers are not very safety conscious" might actually translate into not wearing hard hats or protective goggles on the construction site. Although it is not possible to measure safety consciousness, it *is* possible to count the number of construction welders wearing their hard hats and face shields.

COLLECTING BASELINE DATA

Once the actual target behavior or result is settled on, the behavioral consultant sets out to determine a method to (a) collect baseline data and (b) find the functional variables that affect the target behavior. This is nearly identical to what happens in clinical settings where the behavioral consultant must determine a way to gather the appropriate data in anticipation of a future intervention that will be evaluated. In some PM cases, the data (which are usually important results) can be gathered from existing records. Sales totals, injury records, time sheets, customer complaints, and products returned all generate data. Companies typically keep these types of data in spreadsheets and databases. Increasingly, they are used for accountability purposes and sometimes for goal setting, but rarely are they used in any systematic way to actually change employee performance.

> "Pinpointing leads to observable behaviors, data collection, graphing and analysis of the data, and ideas for interventions."

For example, safety records might indicate how many days were lost because of accidents and what this cost the company, but the actual behaviors that led to the injury will usually escape unnoticed and remain unrecorded. Suggestive selling *behaviors*, which are highly desired these days in restaurants, are rarely actually observed and recorded ("May I offer you some dessert? Our key lime pie is homemade and one of our chef's signature desserts."). Employees go through training, managers are taught to prompt the behavior, and signs might be placed on the wall in the kitchen, but actual direct observation goes just one step too far for most businesses. Most businesses don't know they should record the one thing that would make a difference in the franchise, and the businesses that have a clue that this should be done don't have the skills to do it. This is where the behavioral consultant comes in. Pinpointing leads to observable behaviors, data collection, graphing and analysis of the data, and ideas for interventions. One result that businesses are particularly interested in is the *bottom line*. After expenses are subtracted from income, there is a positive number, a profit, or a *margin*, and this is referred to as the bottom line. A behavioral consultant who can increase the margin of a company will be highly sought after and used repeatedly. Practices will vary from one company to the next, but most of the time, managers and supervisors will be trained in the consulting firm's workshops to observe critical performances. The behavioral consultant (who may be on-site in a company 2 days per week over a 6-month period) will do the training (see Chapter 11) and become the focal point for data gathering, graphing and analysis, and brainstorming about possible cost-effective interventions. Those interventions, as in clinical settings, should be evidence based.

In rehabilitation clinics, residential living, or training organizations, the consultant is unlikely to be working on behaviors that will not have much direct effect on the organization's bottom line. It is much more likely that the consulting behavior analyst will be asked to set up a PM *system* to ensure that behavior programs that are written up and approved are actually carried out precisely, on time, and consistently to produce the desired behavior changes for individual clients. If, for example, Goodwill Industries International hires you to ensure there is follow-up for the training supervisors received on motivating employees with disabilities, the primary target behavior is likely to be "percentage of time the supervisor observed and gave feedback." A secondary target behavior might be "rate of skill acquisition" by the worker. Even though there would be a delay and the margins might be slight when compared with those of competitive Fortune 500 businesses operating on a completely different business model, improvement in these target areas could eventually affect Goodwill's bottom line.

LOOKING FOR FUNCTIONAL VARIABLES

In business settings as well as in rehabilitation or educational settings, pinpointed behaviors are assumed to have a cause. They do *not* just happen out of the blue. Off-task behavior has a cause. There is a reason for it, and it can be found in the *history* of the individual, in the discriminative stimuli (S^Ds) for the behavior, in the motivation system (or lack of), or possibly in the response-cost or schedule of reinforcement. As behavior analysts, we reject the idea that a behavior such as off-task behavior *just occurs* or is due to *traits* in individual employees. Forty years of research in organizational settings tells us that such normal behaviors as coming to work late, failing to clean the restrooms, or engaging in suggestive selling are due to environmental contingencies. The primary job of the behavior analyst consultant is to identify the contingencies and then design a program to correct them. Conducting experimental functional analyses in business or organizational settings is quite difficult, as there is just too much at risk for most companies or organizations to allow these types of manipulations. One strategy is to ask questions about the pinpoint that has been agreed on by management and the behavioral consultant. This will help us find the causal variables. Causal variables are outlined in a brief assessment called the *12 Diagnostic Questions.* The questions are briefly summarized next.*

Antecedents

The first set of questions involves an analysis of the environment to determine if a particular stimulus is setting the occasion for an unwanted behavior or if its absence accounts for a desired or undesired behavior.

Question 1: Does the Person (Employee) Understand What Behavior Is Expected? A related question is "Is there a goal that has been set?" Over the past 15 years, the first author and his students have implemented PM projects of all types with small businesses. In nearly every type of setting, from ice cream franchises to soft drink distributors, we discovered that when employees are asked about detailed specifics of their job, many really don't understand exactly what they are supposed to be doing. The job orientation was incomplete or warped by the person who was assigned to do the training. In the studies conducted with these small businesses, when the supervisor or manager took a person aside and provided direct instruction on exactly how the job should be done ("task clarification"), improvement in performance was about 10% to 30%. As behavior analysts, we understand that improvements

---

* This assessment was developed by the first author as part of the Undergraduate Performance Management Track at Florida State University.

from an intervention as simple as explaining how a job should be done usually don't last long. These results, however, make the point that antecedents in the form of clear job descriptions can make a difference. Many employers forget to set a goal for specific performances, and then they wonder why they don't see the waiters or kitchen staff trying harder. Goal setting can also produce 10% to 20% increases in performance, although, again, this is another intervention that may not sustain results over time. Managers who answered Question 1 in the affirmative can now move on to the next question.

Question 2: Is There a Specific Prompt for the Behavior? An example of a specific prompt is something such as a printed quote attached to the cash register that says, "Would you like a freshly baked oatmeal cookie to go with your sandwich?" A general prompt is something such as this statement made at the end of a monthly meeting: "All right now, listen up. One more thing before we go. We have a memo here from headquarters that says we need to do more suggestive selling. Got it? So, get out there and encourage the customer to consider our add-ons." One thing we've learned from behavior analysis is that antecedent stimuli have to be of just the right type, located in the right place, at the right time. Pilots use checklists just before they take off to make sure they do not miss a single important item related to aircraft worthiness and readiness: "Fuel in the auxiliary tank topped off—check." The last type of antecedent that needs to be considered is the supervisor.

Question 3: Is the Supervisor Present? If So, Does This Individual Provide Any Feedback, Correction, or Reinforcement? Dozens of studies have shown that employees become lax when the supervisor leaves the area. Although this is no surprise to experienced behavior analysts, in many organizations the administrators are shocked to learn this most basic rule. The tendency for work to stop when supervisors disappear is particularly the case if there is no hope for any form of automatic feedback or consequences from the work itself or from the customers. For work that is inherently interesting or where the customers keep the workers on their toes with *their* feedback, the role of the supervisor is much less significant. In the area of developmental disabilities, where employees are supposed to carry out certain kinds of stimulating activities with clients, there can be a very significant decline in such offerings when the supervisor leaves the room. Combine activities that are boring to the staff (e.g., art projects) with clients who provide very little in the way of reinforcement, and you have the formula for an activity that will stop when there is no supervision.

Question 4: Does the Employee Have a Personal Problem That Requires Counseling or Clinical Treatment? Some instances of poor employee performance are due to adverse circumstances that the person has to deal with in his or her personal life. Employees might be chronically late to work for any number of reasons. Sick children or spouses and unreliable cars that need expensive, unaffordable major engine repairs are just a few examples of problems in an employee's personal life that can affect work attendance and performance. In other cases, the employee might be battling depression, abuse, alcoholism, or drug problems at home. The wise behavioral consultant will consider the variables in Question 4 and determine whether they are operating before attempting to set up economic incentives or consequences for poor performance.* To address an employee's personal issues, the behavior analyst will need to work closely with human resources, because many of these matters are confidential.

Equipment and Environmental Variables

For some types of work, the tools and equipment that an employee uses can greatly affect performance; likewise the physical environment is a major factor for certain behaviors.

---

* We are suggesting not that the behavior analyst consultant should try to provide the counseling or therapy but rather that she should refer the employee to human resources for assistance.

Question 5: Does the Equipment Work? Is It in Good Repair? Is the Environment Conducive to High Performance? A few years ago, the first author, along with some graduate students in behavior analysis, was working in the Department of Revenue for state government. The behavioral consultants were trying to track down the controlling variables for "poor response times on phone calls from constituents." The work setting was a room about 30 × 50 feet in size, and it was tightly packed with cubicles. Phones were ringing constantly, and the din was unbearable. A worker who took a call had to hold one hand over the opposite ear to block the noise while trying to take notes with the other hand. The job looked awkward, unpleasant, and painful. The typical expression on the face of the tax advisors confirmed what the workers told us privately: This job was horrible. These employees desperately needed soundproofed cubicles and headsets to do a reasonable job of handling complex questions about corporate tax rates, discounts for the purchase of new equipment, and amortizable losses over 3 years versus 5 years. They knew what they were supposed to do, they knew how to do it, and they had very clear goals, but it was just impossible in this chaotic environment.

Process Analysis

PM was originally derived from the basic methodology of applied behavior analysis that emphasizes understanding the contingencies controlling the behavior of individuals. The methodology handed down from B. F. Skinner gave us a way to focus on the *rate* of responding as the primary dependent variable. For many years, PM carried on this tradition, and the early published work in the field clearly shows how behavioral principles can be used to modify the behavior of workers using basic reversal or multiple baseline designs to demonstrate experimental control. In the 1980s, another approach, *The Deming Management Method* (Walton, 1986), came on the scene from a completely different direction and from another country, Japan. Statistical analysis of error rates, which spawned an approach called "statistical process control" or SPC, has evolved into the still-popular Six Sigma (Pande, Neuman, & Cavanagh, 2000). In behavioral terms, process analysis is a way of looking at chains of behavior rather than at rates of individual behaviors. Behavior chains are an important issue in most organizations,

> "Process analysis seeks to examine the elemental parts of the chain to determine if they are maximally efficient and error free."

and they can involve multiple workers in several departments. Process analysis seeks to examine the elemental parts of the chain to determine if they are maximally efficient and error free. In some cases, this analysis results in a different way of producing a product by eliminating some of the steps or otherwise streamlining the process. Key elements to process analysis include (a) measuring the quality of the product produced at each step, and then (b) gradually changing the way the step is executed using statistical sampling to ensure that quality is continually improved. Behavior analysts working in business, industry, and organizational settings, then, need to be prepared to ask the following questions.

Question 6: Is the Task Designed to Be Carried Out in an Efficient Fashion? Can It Be Streamlined or Eliminated? These questions mean taking a problem that has been presented, such as "We just can't reduce the number of rejects, and our rework is killing us," and translating it into a review and close examination of the process by which the assembly works. It is not uncommon to discover redundant steps or weak links in the process where changes need to be made. This process will involve changing the behavior of individual employees by using interventions such as more training or immediate feedback. This might also involve very frequent direct observation of performance and, under some circumstances, videotaping so that all of the steps in the task can be observed carefully.

Training

Traditionally the solution to every performance problem in business and industry was training. When people did not perform their duties, it was obvious that they needed to relearn the task. Although PM does not ignore training, it recognizes that training is only part of the solution and in and of itself is probably not sufficient. Data from numerous studies over the past 20 years have shown that training alone can have a short-term effect, but overall it will take many other interventions to make lasting changes in behavior.

Question 7: Have You Actually Observed the Behavior? Can the Employee Actually Do the Skill? For many owners or managers, it is assumed that once a person has been through training, performance problems are over. They rarely observe a person before and after training to determine if any skill was acquired. In many cases, the changes in behavior as a result of training are so slight as to be negligible and invisible to the naked eye. By asking the referring person about the effects of training, you are essentially questioning the value of the training and suggesting that it might need to be examined as a variable and perhaps replaced with something more effective. If the answer to this question about whether training is effective is "no," then the PM consultant will want to set up a systematic evaluation of the training. Depending on the results, it is likely that a complete revamping of the training methods will follow to make training more performance or outcome oriented.

Contingencies of Reinforcement

Questions 8 through 12 are about consequences in the environment that might affect an individual's performance. Behavior analysts recognize how powerful these variables are, but because few people in the business world seem to understand this, a series of questions are designed to bring out the necessary information.

Question 8: Does the Behavior Produce an Observable Effect? By asking this question, you will be required to determine exactly what happens when an employee engages in the desired behavior. In some work settings, the behavior produces no effect whatsoever. Many tasks require employees to put in a great deal of time before they produce anything, and in this computer age, people toil at keyboards on tasks that appear to fly off into the network. Not only are there no effects and no products but no one notices. The task of the behavior analyst is to determine if the environment can be changed to add feedback to the system. To have the maximum effect, immediate feedback is preferred over delayed feedback.

The first author, along with his PM students, worked on a project at the Bureau of Motor Vehicles (BMV). In one particular BMV job, hundreds of data entry clerks filled out forms by the thousands every day. The opportunities for errors were many, but the typists got feedback only once a month in the form of a big spreadsheet. After inquiry, the consulting behavior analysts learned that the computer terminals could be programmed to give feedback hourly. An administrative decision had been made that this was not necessary, so this feature was not built into the computers used by the data entry clerks. The consultants were told, "People who make a lot of errors are fired or sent back to HR for more training." What an extreme—errors could result in more training or being fired!

Question 9: "Is a Competing Behavior Being Reinforced?" This question examines an issue that will be overlooked by most people who are not behavior analysts. We understand that individuals can engage in multiple behaviors that occur very close in time and those behaviors that are reinforced are likely to occur again. A person assigned to greet customers with, "May I help you?" needs to keep an eye on the entrance to the shop and remain in motion, always vigilant for new customers. At the same time, if the greeter's colleague is standing beside the greeter telling a juicy story about partying in Cancún, for the greeter, engaging in listening behavior is likely to be much more reinforcing than watching for new customers. If you discover that a competing behavior *is* being reinforced, as a behavioral consultant you will need to redesign the contingencies to reduce the competition and/or increase the reinforcement for the desired performance. Managers who don't understand subtle contingencies of reinforcement try to solve problems like this with specific requests, demands, and vague threats, "Some of you, and you know who you are, need to be paying closer attention to customers at the door or the door is likely to hit you on your way out." In reality, managers and supervisors need to realize that the reinforcer greeters get from greeting customers simply doesn't compete with a juicy story from a college-age colleague who just came back from spring break in Cancún. Chatting with friends instead of working can cost a restaurant business and result in ill-will toward an entire franchise. Supervisors need to find stronger reinforcers to compete with off-task behaviors and if there is a problem, they will likely have to spend time in the front of the restaurant prompting and reinforcing more customer-friendly behavior.

Question 10: "Is There Any Response Cost or Other Punisher Associated With Performing the Task?" In some cases the primary reason that a desired behavior is not occurring is that it has a built-in, response cost or response effort that serves as a sort of automatic micro-punisher. Waiters and waitresses who are required by restaurant management to engage in up-selling or suggestive selling often find such tasks aversive because of the responses they elicit. Customers occasionally make sarcastic, demeaning comments, "If I wanted wine, don't you think I would have ordered wine?" or "Do I look like I should be 'considering dessert'?" Such punishing consequences make suggestive selling much less likely to occur with restaurant wait staff. Further, if this is linked with the detail that there is no reinforcement if the customer says "yes," it is easy to see why this behavior does not occur. If as the PM consultant, you could convince management to provide some sort of tangible reinforcer for suggestive selling, e.g., fifty-cents for each dessert sold, this might offset the aversive nature of the task to a certain extent. Other forms of response cost or response effort are a little more difficult to identify. At one consulting firm, the CEO was having trouble getting the field-based consultants to turn in their billable hours in a timely fashion. After sending email prompt after email prompt to, "download the pdf, fill it in, and just drop it in the mail to me by the 15th," he decided to take a closer look at the process and interviewed two of the most consistently delinquent individuals. They complained of having to gather up scraps of paper, find the form on their desktop, add up all the numbers, calculate the different rates for each client, fill out the form, find a stamp, then an envelope, and take it to the mailbox before the mailman arrived. The CEO took all of this to heart and hired a computer programmer to write a billable hours program that allowed each consultant to enter billable hours each day (raters were automatically calculated). Then, at the end of the billing period, with just a couple of clicks, the hours were sent to headquarters. Our wise, behavior analyst CEO found that he could get 100% on-time compliance just by reducing (almost completely) the response cost with turning in hours. We now know that every behavior we engage in throughout the day has some built-in response cost or response effort. If you want faster responses, higher rates, or better quality behaviors look for these built-in aversive consequences and see if they can be reduced or eliminated.

Question 11: Is There Any Kind of Feedback From Peers, Customers, or Supervisors for the Behavior? This question asks about the culture where the work gets done. We know that simple feedback on a behavior can make a big difference in the future likelihood that the response will occur again. If customers reinforced wait staff for their excellent service, it would probably be repeated more often. If their peers made positive comments, the desired behaviors would be strengthened. All too often, however, customers make no comments and neither do the worker's peers. The last hope is that managers will provide some sort of feedback, at least occasionally, when they observe desirable behaviors. Sadly, most managers rarely give feedback for completing tasks correctly. Instead, the general tendency is to sit back and zap people when they make a mistake or provide poor customer service. If you are consulting in a company where the complaint is that certain behaviors are missing or occurring at a low rate, you will definitely want to spend some time observing to see what the rate of positive feedback is for desired performance. Don't be surprised if you see none at all. By asking this question and taking notes on supervisor behavior, you will begin to develop a strategy that could make a significant difference in this work environment. Note to self: "Need to work with the supervisor and teach him to observe more, detect desired performance, and give positive feedback."

Question 12: Is There Any Intrinsic or Extrinsic Reinforcement for the Behavior? Our last question has to do with the actual reinforcers that are closely associated with the desired performance. We leave this question for last, because in terms of intervention, it is the most difficult to engineer, and in terms of cost to the organization, it represents possibly the most expensive

> "Some jobs require very little in the way of extrinsic reinforcement because the *intrinsic* reinforcement is so great."

investment. Some jobs require very little in the way of extrinsic reinforcement because the *intrinsic* reinforcement is so great. People who love gardening and enjoy talking to people would probably be highly motivated to give great customer service if they worked at a garden center. They would smile at each person, share their love of plants with each customer, and go to great lengths to find just the right perennials for the customer's yard and garden circumstance. This type of matching a person to a job represents an ideal situation, because most jobs are a total mismatch. Imagine the same job, but now picture a brooding, socially awkward teenager whose parents required that he apply for and accept this job for the summer. Chances are the teenager will attempt to avoid customers and provide them with the least amount of information possible. This teenager, who was forced to do a job he didn't care about, might sneak around the corner of the potting shed and have a smoke. He might provide a new reason for calling it a *potting* shed, or he might spend a lot of time on his cell phone, texting or chatting with friends. The intrinsic motivation found when in working in a garden center is totally lost on this young man. To get him to the point of providing a high level of customer service would require a lot of training, supervision, and *extrinsic* motivation, probably in the form of money. For the more ordinary workplace where most of the work gets done on computers and there is little in the way of direct contact with customers, there is probably zero intrinsic motivation, because the task itself provides no satisfaction. In a situation like this, it is likely that boredom will quickly set in. No amount of training can overcome the built-in response cost and fatigue that comes with an unstimulating task; management will have to find ways of providing extrinsic motivation for these workers.

SUMMARY

The process of PM consulting is very similar to the consulting process in schools or with families. You begin with a referral, work with your contact person (in the setting) on pinpointing the problem, and then do direct observations and informal data collection. In the business environment, you will next need to submit a proposal indicating the scope of the work you plan to do along with the estimated time and cost to complete the work. Contracts in business and industry are generally in quarterly intervals, that is, you will submit a time and work proposal for a 3-month or 6-month period. PM consultants usually work for firms that have divisions or specified staff that prepare the contracts. A salesperson will work with the company on the details. If you are working on a smaller scale such as a rehab center that is having a problem of tardiness or turnover, you will more likely submit a proposal to work on only one target behavior. You will submit an estimate of the number of hours you think this will take. Just as with clinical and educational consulting, consultants in PM consulting rely on mediators to carry out their designs, so you will need to build in time to do training of the personnel necessary for you to accomplish your task.

The amount of time you'll need can be considerable, because most administrators are not used to thinking about how to behaviorally manage their organizations. They are more comfortable with a command-and-control model where they issue instructions and expect them to be followed. They assume employees will do what they are told, come to work motivated, and not require feedback. You will need to use all your persuasion skills (Chapter 9) in addition to your basic business etiquette (Chapter 1), assertiveness (Chapter 2), and leadership (Chapter 3) repertoire to maximize your success.

FOR FURTHER READING

Abernathy, W. B. (2000). *Managing without supervising.* Memphis, TN: PerfSys Press.

Daniels, A. C., & Daniels, J. E. (2004). *Performance management: Changing behavior that drives organizational effectiveness.* Atlanta, GA: Performance Management.

Pande, P., Neuman, R., & Cavanagh, R. (2000). *The Six Sigma way: How GE, Motorola, and other top companies are honing their performance.* New York: McGraw-Hill.

Rummler, G. A., & Brache, A. P. (1995). *Improving performance: How to manage the white space on the organizational chart.* San Francisco: Jossey-Bass.

Watton, M. (1986). *The Deming management method.* New York: Dodd, Mead & Company.

Section

Four

Vital Work Habits

17

Time Management the Behavioral Way

Take 5 minutes to plan your day.

Brian A. Iwata

Behavior analysts are busy people who are generally in great demand by consumers, administrators, teachers, parents, treatment coordinators, and others looking to improve the life of their clients. Although there is no "average" day, a typical day might start early in the morning with a Board Certified Behavior Analyst® meeting with a principal to iron out some issues with a client who has gone astray. The behavior analyst goes from this meeting to work with a teacher who is having a problem with a new behavior program. In the meantime, the busy behavior analyst's cell phone has vibrated three times, indicating she has phone calls to return as soon as she gets a break. Her break will probably be in her car as she is on the way to a client's house to supervise an intern. And so it goes throughout the week, with the behavior analyst rushing to avoid being late because traffic was heavy or a client was having a meltdown. The requirements of the job also mean keeping track of billable hours in quarter-hour intervals, completing progress reports, and making presentations to teams developing individualized education programs.

Learning to plan your day and manage your time is an essential skill for the effective behavioral consultant, and there are many,

many distractions that will be competing for your attention. As you reflect at the end of an average day, you realize your 30-minute meeting with the principal went well because you were able to give her the advice she needed on how to handle an aide who is unresponsive to feedback; in one of the phone calls from a former client, you were able to provide some reassurance he made the right choice of a girlfriend and job placement; and you regrettably had to turn down an e-mail invitation from a friend who invited you to meet with her for drinks after work. As much fun as it would have been, this was bad timing because you have a big report due the next day pertaining to a situation for which you could end up in court.

HOW TO WASTE TIME

As a change of pace, let's take a look at a few ways that you can waste your most precious resource: time.

First, the most obvious way to waste time is to do tasks that others can and should be doing. The Board Certified Assistant Behavior Analyst® (BCaBA) who is supposed to collect and graph data for presentation at a meeting shows up late (saying she really didn't understand the instructions), so you spend 25 minutes prepping the data for a meeting at 1:00 p.m. This keeps you from observing a functional analysis session with a self-injurious behavior (SIB) client. Now you have to reschedule the functional analysis (FA) session for Monday, and Monday's schedule is already full. The solution, of course, is to make sure that your instructions to your BCaBA are clear in the first place. Did you give her plenty of examples? Did you have her practice one while you watched? Did you provide corrective feedback? Did you tell her where to go if she had questions? Did you emphasize the importance of completing the task on time?

Second, if you analyze how you spend your time each day, you may discover that you have allowed behavioral erosion or slippage to take place. Meetings that you run or attend that are supposed to be 1 hour long are taking an extra 15 minutes because of poor meeting management. If you are in charge, can you scrutinize your agenda items to determine if some can be deleted? Rather than taking up 15 minutes at the beginning of the meeting with announcements, can you just print those and hand them out? Are certain people dominating the meeting with their inane comments? And what about your written work? Without compromising quality, can you streamline your writing so you can crank out your reports in 30 minutes instead of 45? Is there a way to keep track of your billable hours on your PDA as you complete each hour rather than rummaging through crumpled notes at the end of the week, trying to figure out where you went and what you did?

Third, having to rework projects can be a waste of your time. If you look at what you've done during the week and realize that some of it involves having to redo an assessment report because you left out the reinforcer preference scale data or another similar detail, maybe it's time to think about a task analysis of your work output. Eliminating rework could amount to saving you as much as an hour per week.

Finally, allowing others to take up your time with trivialities is wasteful. This is a touchy subject for some behavior analysts because it appears to conflict with networking (Chapter 4) and becoming a trusted professional. Most behavior analysts, by their very

> "Most behavior analysts, by their very nature of being positively reinforcing people, will be sought out as friends, advisors, and confidants."

nature of being positively reinforcing people, will be sought out as friends, advisors, and confidants. Although this is important, and certainly gratifying, it can take up valuable time that you should spend getting your own work done. The key cautionary word in this is *trivial*, which has different meanings to each person. A friend is in the hospital having a baby; do you *really* need to drive across town to see her? Of course you do, as she is your best friend. A neighbor invited you to a birthday party for her 7-year-old niece on the weekend. Do you have to attend? Probably not, but if your work is not done and you feel that you want to attend the party, consider staying a shorter time and getting your work done for Monday. It is football weekend, and everyone will be tailgating and going to the game. If you feel like you just can't miss this activity, get your work done in advance. Learn to say, "I'll be busy all weekend, so I can't go out this week. Thanks for inviting me to dinner, but I need to stay home and get everything done so I can go to the game and parties this weekend." Examples of people inviting you to participate in activities that can distract you from your work are endless. Ideally, you'll be well rounded and will find the healthy balance between enjoying your leisure time and meeting all of your professional obligations. Our general advice is for you to be kind, caring, and analytical in your decision making regarding nonwork requests of your time while managing your schedule in such a way that you get your work done and maintain your professional reputation.

GETTING THINGS DONE*

David Allen, with his careful and detailed analysis of the daily stumbling blocks that stand in the way of accomplishing important outcomes and his insightful suggestions on how to cut through the maze with a few rules to simplify and organize your work life, has become the guru of productivity (Allen, 2001). As shown in his workflow diagram (see Figure 17.1), he recommended making a series of decisions about all the "stuff" that comes across your desk (or computer screen).

First, you need to decide if what is in front of you requires some action on your part; if not, you can trash it or file it. If it requires action, Allen suggested asking, "Will it take less than 2 minutes?" If "yes," then do it. If not, you can decide to either delegate the

---

* This is the title of a best seller by David Allen that everyone should read. He also has a Web site that is very useful for getting organized and improving productivity: www.davidco.com.

WORKFLOW DIAGRAM—ORGANIZING

Figure 17.1 David Allen's workflow diagram. From *Getting things done: The art of stress-free productivity,* Allen, D. (2001). New York: Penguin, p. 139.

task or defer it to some other time. If you choose to defer the task to a later time, either put it on your calendar for a specific time or put it in a "as soon as I can" file. Allen has created several conceptual and pragmatic tools that, when used together, greatly increase the likelihood that you will be productive. The "2-minute" rule, for example, is great for deciding what to do now and what to do later, especially because if you are going to do it later, you have to decide to add in a second set of "buckets." Note that asking the delegation question is important in his productivity system. In his experience, asking, "Am I the best person to be doing it?" (Allen, 2001, p. 133) can greatly reduce your workload and increase your effectiveness. If the item in front of you will take longer than 2 minutes and you can't delegate it, it goes into the "Next Action's" bucket. This is similar to a to-do list, but it is, however, a new and improved one that recognizes that larger "Projects" have many steps to be completed before they are actually done. One final innovation that Allen included in his system is the Weekly Review. By requiring yourself to look at all of your current Projects and Next Actions as well as "Waiting For" items (this includes delegated work assignments), you are essentially eliminating the possibility of having a project fall through the cracks of your busy workweek.

TIME MANAGEMENT THE BEHAVIORAL WAY

Allen's system covers all the important aspects of time and project management, and we highly recommend his book to all new behavioral consultants. There is one feature that is not included that may be important to behavior analysts who are interested in the motivational aspects of work completion. Allen assumed that task completion itself is a natural reinforcer. It may be for many people, but for some there is a need for something else, such as a consequence to keep moving ahead with tasks that are boring, arduous, or unfulfilling. To fill the gap we suggest a basic behavioral procedure with which you are no doubt already acquainted: the Premack Principle. The Premack Principle basically states that more probable behaviors will reinforce less probable behaviors. For example, we assume that you have a list of things you'd like to do each day, from playing a video game on your cell phone to checking out a YouTube clip, texting or calling a friend, renting a DVD, or going out for pizza with your colleagues. For many people, these are tasks for the end of the day when they are ready to relax. But for other somewhat undisciplined people, these preferred tasks cause trouble when they stand in the way of work or chores that need to be done.

Another way to conceive of preferred activities is as "high-probability behaviors" that can be used as reinforcers for the completion of certain tasks that are not naturally reinforcing. Five minutes of checking out YouTube could be made contingent on returning a bunch of work-related phone calls that need to be made but are just not that much fun. Going to lunch with a friend would be a lot more enjoyable if you arranged for the lunch to follow turning in a report that has been due for several weeks.

This concept adds another level of planning to your daily schedule. In addition to making a list of the tasks to be completed, consider the Premack Principle and develop an order in which you will complete the tasks (see Figure 17.2). Using the behavioral approach to develop the new, high-powered, productive you, you will translate your "things to do" into behaviors. You'll have a behavior, and following each behavior, you'll have the Next Action.

To spice up your day, make it more interesting, and, of course, more reinforcing as you fill out your Next Actions list, consider

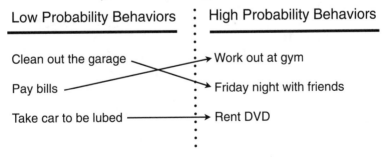

My Self-Management Plan

Low Probability Behaviors	High Probability Behaviors
Clean out the garage	Work out at gym
Pay bills	Friday night with friends
Take car to be lubed	Rent DVD

Figure 17.2 An example of using the Premack Principle to reinforce daily work activities.

adding an occasional high-probability behavior as a consequence. We also recommend the simple act of putting a distinctive check mark by each item completed so that at the end of the day you can visually scan your calendar and see how much you accomplished. If you haven't converted to an electronic calendar, such as iCal for the Mac or Google Calendar for your BlackBerry, we recommend a simple paper-based planner, the Day-Timer (see Figure 17.3).*

TIME MANAGEMENT VIA BEHAVIOR MANAGEMENT

If you recognize that time is a very valuable resource and should be savored during only the most important events, you will appreciate this last tip. Every day, you will no doubt be approached with opportunities to take on responsibilities (someone is trying to delegate to you), share experiences, or volunteer your time for a worthy cause. As a behavior analyst, you are a likely target for these offers because you are probably a reasonable, effective person who appears

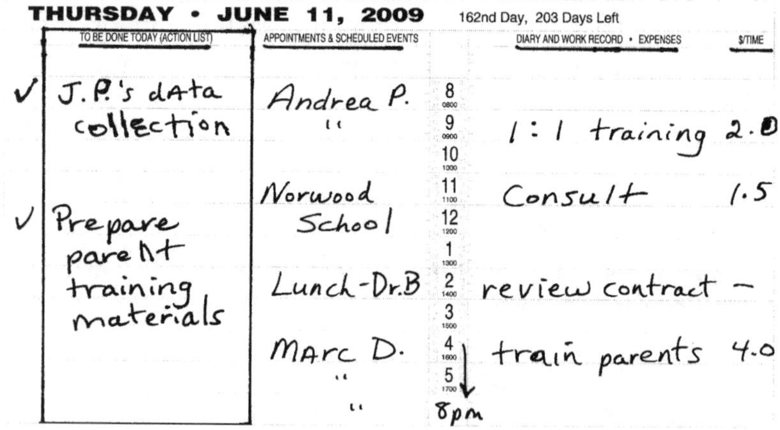

Figure 17.3 A page from the Day-Timer planner.

---

* The Day-Timer has a wide variety of products to help you become more productive. For more information, go to the Web page at www.daytimer.com.

to be congenial and open to others. And at the beginning of your career, it is exciting to be given so many options to engage in interesting and worthwhile tasks with friends, relatives, and col-

> "If your long-term goal … is upward mobility in your company or community, you will need to (a) discriminate those offers that have relevance and worth from those that are less important and (b) learn to say 'no.'"

leagues. If your long-term goal, however, is upward mobility in your company or community, you will need to (a) discriminate those offers that have relevance and worth from those that are less important and (b) learn to say "no." You have to engage in serious considerations when deciding to turn down an offer. A big concern is offending the person extending the offer.

HONESTY IS THE BEST POLICY

You can make matters worse with a flimsy excuse that is transparently false. Your goal in saying "no" is to let the inviter know you appreciate the invitation, but you need to firmly express that you are just unable to accommodate the invitation at this time. It is a very bad idea to make up an excuse: "I'm sorry, I can't. I have to go to a funeral." This might get you off the hook at the moment, but if the person sends you a sympathy card and later sees you at the mall or a movie when you were supposed to be at the cemetery, you are in serious trouble.

There will be pushy people who will invariably respond to your polite "No, I'm sorry, I can't. But thank you for the invitation" with "Why not?" Remember, you do not need to provide an explanation, and we would stay away from excuses. It is probably best just to assume a firm posture, make good eye contact, and repeat your original reply with a firm, objective tone of voice: "I'm sorry, I can't. But thank you for the invitation."

SUMMARY

Because most jobs in behavior analysis will involve many, many small tasks that will vary during each week, time management is a skill that can make or break the consulting behavior analyst. Planning your day, avoiding wasting time, using Allen's method for handling workflow, and applying the Premack Principle in your own life will help you develop into a person who does high-quality work and turns it in on time with energy and enthusiasm to spare.

FOR FURTHER READING

Allen, D. (2001). *Getting things done: The art of stress-free productivity.* New York: Penguin.

18

Become a Trusted Professional

Trust is my faith in your ability or word in some specific area.

Hyler Bracey (2002)

An enthusiastic first-year graduate student in an applied behavior analysis program was looking forward to starting her internship and learning new skills. She was looking forward to meeting her clinical supervisor, who had excellent behavioral skills and who had promised to teach her everything she needed to know to be a good behavior analyst. Full of excitement, the student went to her first scheduled meeting with the clinical supervisor. The supervisor started the meeting by telling the student she needed to adjust her schedule of school visits to match the supervisor's. The student was informed she would receive training, observations by the supervisor, and feedback once a week, starting on the following Monday. This was great news. After all, the supervisor had 15 years of experience and came with a reputation as a behavioral clinician who could solve nearly any problem. When Monday came, the supervisor rushed into the room more than 30 minutes late. He blurted out, "I really don't have time to work with you today. Just show me what you know how to do." The student was observed, but the supervisor was reading and returning e-mail on his cell phone the whole time. At the end of the session, the supervisor said, "You're doing fine," and that was it. The following week, the supervisor called the student 5 minutes prior to their appointment and said he was canceling, muttering that his wife was out of town and he had to "take Jenny to the doctor, she's running a fever." The following week the supervisor did not show up and did not bother to call. The student called the supervisor and left messages that were not returned. Two more weeks went by with no contact from the supervisor. The semester was now half over, and the student had received no training and no feedback. In the eighth week, the supervisor showed up on time, said, "How's it going?" and proceeded to make two calls on his cell phone. He didn't apologize for being absent or offer any explanation. When he left the building, he said, "OK, see you next week" and hurried off.

Unfortunately, this scenario is a true story involving a supervisor who appeared to do everything he could to actively build a lack of trust. He may have had very good reasons for missing sessions or being late, but the student would rightfully be justified in having very little trust in this supervisor going forward.

As a behavior analyst, you will find yourself interacting with behavioral and nonbehavioral colleagues every week on personal and professional issues of every stripe and kind. They will offer advice, counsel, and admonitions about a wide variety of topics that affect you and your clients. Over time, you will come to respect some of these individuals for their honesty and integrity. Others, you will put on a lower tier as being unreliable and untrustworthy and perhaps having questionable character. Likewise, you will be judged by others on the basis of your behavior toward

> "If you are a good listener, provide thoughtful suggestions, and go beyond the call of duty to help those in need, you will be seen as a valued member of your profession, and *you* will be trusted."

them, the advice you give, and the positions you take on professional issues as they arise. If you seem aloof and indifferent to the suffering of others or make light of their plight, you will be judged harshly. In contrast, if you are a good listener, provide thoughtful suggestions, and go beyond the call of duty to help those in need, you will be seen as a valued member of your profession, and *you* will be trusted.

ACHIEVING TRUST

Trust is hard to achieve and comes only through hard work over time. To earn the trust of others, you need to demonstrate steady, consistent temperament and be honest and reliable (Harvard Business School Press, 2005). "Trust me" is perhaps the single most overused expression in business today and, of course, is often a dead giveaway that the speaker is certainly *not* to be trusted with anything. True trust is earned over time as you engage in your daily activities with colleagues and clients. You promise to meet a person at a certain time, and, despite the horrific weather, you show up. You listen to the plight of a friend who shares a confidence, and you *never* reveal it to anyone. You express your strong support of evidence-based treatments, and despite pressure and pleadings from a client, you refuse to endorse a fad treatment. All of these are examples of ways in which trust is gained over time.

It seems that in every office, there is a jokester who will say anything to get a rise out of people. The jokester's inappropriate comments, which can be crude or insensitive, are usually followed by "Just kidding." Individuals who engage in such antics may be amusing, and they might help lighten up the day in the workplace, but they are unlikely to be trusted by those around them. People who can't get the fooling around under control are often considered gossips, cranks, pranksters, and lightweights, and their occasional expressions of commitment and sincerity are not taken seriously. A related concern for the behavior analyst who wants to become a trusted professional is the old adage "You are known by the company you keep," which is clearly relevant here.

DEVELOPING TRUST

It is important for behavior analysts to become trusted professionals so their ideas on important matters are sought out and their opinions are respected. This status is not conferred automatically because of a title or credentials. Trusted status is earned in small increments over time. You will have an opportunity to build trust every time you participate

> "You do not need to offer an opinion on every topic or push your ideas at every meeting ... being somewhat reserved and perhaps reluctant to be drawn into the fray is one sign of a person who wants to be taken seriously."

in a meeting, help a client make a decision about treatment, offer an opinion about a hot topic, or make a commitment, even a small one. You do not need to offer an opinion on every topic or push your ideas at every meeting; in fact, being somewhat reserved and perhaps reluctant to be drawn into the fray is one sign of a person who wants to be taken seriously. Thoughtfully considering the issues and taking the time to deliberate and research a topic *before* expressing an opinion are hallmarks of someone we trust. A person who is bubbling over with half-baked thoughts on every topic seems to be someone who is primarily looking for attention rather than someone who can be trusted to offer a sound alternative.

GETTING STARTED BUILDING TRUST

The first step to *becoming* a trusted professional is to be able to identify the characteristics of these rare individuals. A trusted professional is a person who, above all else, is honest in her dealing with others, is fair in her assessment of difficult situations (i.e., unbiased, not judgmental), and does not place blame but rather uses her behavior analytic skills to find workable solutions that are fair to all parties. In behavioral terms, we say a trusted professional is a person who responds to the available evidence and is consistent in that regard. Subjectively, the trusted individual is someone who is steady, calm, reserved, deliberate, thoughtful, consistent, reliable, confidential, dependable, loyal, and steadfast. If you look around and ask, "Who are my colleagues who have these characteristics?" you will be headed in the right direction. Once you identify a person who is trustworthy, you will want to seek that person out, begin to observe the person in various situations, and, if possible, associate yourself with this professional. It is likely that these trusted professionals are in a leadership position, have considerable visibility, and are no doubt busy. You don't want to become a pest or insult them with flattery; quietly observing how they handle difficult situations is your first best strategy. If you can be part of a working group with this person and learn firsthand how this individual looks at the world and sizes up issues as they arise, you will learn a great deal.

TRUST

In his great little book *Building Trust* (Bracey, 2002), Bracey outlined five steps spelling out TRUST. He delineated the following steps a professional can take to build trust:

Be **T**ransparent
Be **R**esponsive
Use Caring
Be **S**incere, and
Be **T**rustworthy

Be Transparent

Bracey argued that for others to trust you, they have to be able to see how you think through issues. You also have to be "easily readable" to those around you. If you come up with a recommendation

totally out of the blue, not connected with previous ideas or
suggestions, it may be difficult
for others to trust your logic
or your judgment. Being able
to logically think aloud about
a problem so that listeners can
tell that each step is logical and
sensible is a great way to build
others' trust in your judgments.

> "Being able to logically think aloud about a problem so that listeners can tell that each step is logical and sensible is a great way to build others' trust in your judgments."

You won't have to do this with all of your ideas, but it certainly
helps in the beginning when you are trying to establish yourself as
trustworthy.

A second method of establishing trust involves presenting
"easily readable behavior" when you are dealing with people
(Bracey, 2002, p. 20). Bracey said that allowing people to see
how you feel about situations builds trust in the sense that they
know where you stand and will not be surprised by some decision down the road. Tactfully letting people know that you are
pleased or unhappy with how a project is going gives them the
information that they need to make corrections. They will respect
you and trust you, especially if your decisions match your readable behaviors. The poker-faced administrator makes everyone
uneasy because people never know where they stand with this
person. Trust comes from projecting strategic openness during
critical meetings.

Be Responsive

Bracey asserted that trust also comes from being responsive
to those around you, that is, giving feedback in a constructive,
spontaneous, and caring manner (Bracey, 2002, p. 23). We usually think of feedback as a way of changing someone's behavior,
but Bracey put a different twist on this, suggesting if the purpose
is to help the other person, the end result will be that the individual will come to trust you. Giving positive feedback changes

behavior and builds trust. Thinking more broadly about the reason we give feedback allows us to see that Bracey's observation is quite true. By shaping on someone's behavior, we are essentially saying, "I am prepared to make an investment in your future. I know that you have potential, I see that you are trying, and I am prepared to make you successful. You are worthy of feedback, and I am sure that you can improve and make a difference in the life of these clients." As behavior analysts, we often miss this interpretation of the importance of giving positive feedback to those around us. In doing so, we may have also missed the opportunity to build trust.

There will be times when negative feedback is necessary. Negative feedback should be used in an unemotional, constructive fashion to let the learner know he or she did not get the behavior quite right: "Thanks for turning your report in on time, Jim. I need to ask you to do one part of it over and include a graph that shows changes in the target behavior this month." Negative feedback will often be dismissed if it has not been preceded by healthy doses of positive reinforcement. In fact, people who receive positive feedback as well as negative feedback will probably trust those supervisors more than those supervisors who never attempt shaping and make only positive comments.

Use Caring

Gaining the trust of colleagues means paying attention to the subtleties of social interactions. The way in which you react to their questions, comments, or presentations can make a big difference in whether your colleagues will trust you with certain kinds of information. Making eye contact when listening, paraphrasing what they say, and letting them speak without interrupting are all ways of acknowledging other people and building trust. Being careful with your language and not putting another person on the spot if his peers or supervisors are close by is an important aspect of building trust. Saying, "I'm confused. Can you go over that again?" is much better than saying, "You're confused. I have no idea what you're talking about." Dale Carnegie addressed this issue when he said, "Let the other person save face" (Carnegie, 1981). Using this strategy is especially important if the person *is* confused, because pointing this out publicly could do a lot of damage to your trust factor. Taking the time to choose your words carefully and let others know you care about their feelings will pay off when the time comes when you need someone to go the extra mile. When asked, people who trust you will come through for you because they know that you truly appreciate their efforts.

Be Sincere

Being able to match what you show in your facial expressions and body language with what you say and what you do is the formula for creating a sincere and trusting relationship with those around you.

> "Any efforts you make to develop trust will likely fail if you come across as insincere."

Any efforts you make to develop trust will likely fail if you come across as insincere. Using obvious flattery, smiling when you are not happy, and using flat-affect positive reinforcers such as "Good job" are detectable by friends and colleagues and will have the reverse effect of convincing others of your sincerity. Todd Risley (1937–2007) was known as a pioneer and genius in the field of behavior analysis. Among his many contributions to the field was his work on Say–Do congruence (Risley & Hart, 1968). In Risley's research on early childhood, he showed ways of teaching preschoolers to tell the truth by shaping on their Say–Do behaviors. As adults, we can use the same sort of contingency on ourselves to show others they can trust us to show good judgment and make good decisions. If you are trusted, your colleagues will listen to what you have to say and follow your lead.

Be Trustworthy

Being seen by others as a trusted professional has a certain downside. If you agree to do something and fail to do so, there will be consequences. Your reputation will be damaged, and you will lose a degree of trust that will have to be regained at some future time. Once trust is lost, recovery can take a while, and the process is painful. Understanding that behaviors have consequences is not new to behavior analysts, although we are more inclined to think about this in terms of other people's behavior rather than our own. Paying attention to the small commitments that are offered every day can make a big difference in whether you are trustworthy in another person's eyes. Someone saying, "Why don't you join us for drinks after work?" is a compliment of sorts. The inviter would like to spend some after-hours time with you. When you respond with, "Sure, where?" it sounds like you'll show up. If you have no intention of stopping by for happy hour but just didn't know how to say "no," you just squandered a little bit of your credibility and trustworthiness. Do this enough, and you'll get a reputation of being a person who is unreliable. A socially sensitive person might even say you lied to her or embarrassed her because she told her friends you'd be there. On the flip side is the behavior analyst who is looking to build credibility and trust by forcing herself to make small commitments to do something at a specific time. It actually doesn't seem to matter what the commitment is as long as you follow through. A minor commitment is "I'll send you a PDF of that article. I have it on my computer at home and will send it by 8:00 p.m." In the big scheme of things, it might not matter if the person gets it tonight, tomorrow, or next week. It absolutely does matter, however, in terms of building trust. It's likely that the recipient will be somewhat surprised to receive the document right on time. If you can do this sort of trust-building exercise with all the important people in your life, both personal and professional, you'll find that they will come to see you differently from everyone else they know. Note that in this example, you could have *not* promised the PDF by 8:00 p.m. and just sent it as a courtesy, but to do so does not build trust. The gain in trustworthiness comes from the Say–Do contingency. Building trust in this way involves only two rules: First, do not make any agreement that you don't intend to follow through with. Second, go out of your way to make frequent small commitments that you can and will honor so as to programmatically build trust in colleagues, friends, supervisors, and clients.

BECOMING A TRUSTED PROFESSIONAL IS ESSENTIAL FOR BEHAVIOR ANALYSTS

The behavioral paradigm is foreign to many and contrary to that of most other professionals who work in the human services arena. A child who is disruptive in his second-grade classroom (and does so to gain attention from the teacher and peers) might have a behavior program that includes extinction for tantrums.* This idea is certainly contrary to what the school counselor, school psychologist, and assistant principal would likely advise. The common thinking on this type of behavior is that the child needs attention ("He just needs to come down to my office and talk to me about his problem."), testing, or perhaps discipline ("He just needs to come down to my office for a good talking to and maybe a few hours of in-school suspension."). If the behavior analyst consulting at this school is going to get any buy in, when it comes to working with the teacher and other professionals, she will need to have some trust in the bank. If the behavior analyst isn't seen as a trusted professional, no one will support the program she is presenting that involves ignoring minor disruptive behaviors, especially the teacher who is on the front line in this battle for control of her classroom. Furthermore, if the

---

* Assume that the Board Certified Behavior Analyst® has done a proper functional analysis and determined this is an attention-maintained behavior.

behavior analyst asks the teacher, who does not like this student one bit, to consistently use differential reinforcement of other behaviors (DRO) with little Bobby, she is likely to get a response such as, "Give him praise for sitting quietly? After what he just said to me? I don't think so!"

One strategy for gaining trust is to quickly solve some simple problems for key people who need help. Demonstrating that you are effective will help considerably with the trust issue. What others never appreciate are a lot of excuses for why you can't help them, why a program isn't working, and why you are so hard to reach when the chips are down. Building credibility and trust every single day will result in the cooperation and support you need to implement effective programs.

SUMMARY

Effective behavior analysts gain the trust of others by being good listeners, providing thoughtful suggestions, keeping confidences, and going above and beyond what is expected to help others. Trusted professionals are known as honest, reliable people who give thoughtful consideration to issues before replying, and they can be counted on to do what they say they are going to do.

FOR FURTHER READING

Bracey, H. (2002). *Building trust: How to get it! How to keep it!* Taylorsville, GA: HB Artworks.

Carnegie, D. (1981). *How to win friends and influence people.* New York: Simon & Schuster.

Harvard Business School Press. (2005). *Power, influence, and persuasion.* Boston: Author.

Risley, T. R., & Hart, B. (1968). Developing correspondence between the non-verbal and verbal behavior of preschool children. *Journal of Applied Behavior Analysis, 1*(4), 267–281.

19

Learn to Deal Behaviorally With Stress

Stress is the harmful physical and emotional responses that occur when the requirements of the job do not match the capabilities, resources, or needs of the worker.

National Institute for Occupational Safety and Health Administration (1999)

Behavior analysts are particularly prone to stress by the very nature of their work. They withstand long hours in direct physical contact with clients who are unpredictable. Clients can, and do, bite, scratch, kick, or punch when it is least expected. Adding to the stress factor is the fact that behavior analysts almost always work in the public eye, where they are open to criticism by those who are not well versed in our theory or methodology or are less committed, poorly trained, or lacking any background in human behavior.

Driving across town or from one county to another to reach clients who live in remote areas can add further stress to a behavior analyst with an already hectic schedule. Enduring traffic delays, receiving urgent cell phone calls while you are negotiating a detour, or spilling your hot coffee while making a U-turn can turn the commute from one client to the next into a nightmare. Parents often have an expectation that behavior change will come quickly, and they may act disappointed if little Adam is not cured of his hyperactivity, aggressiveness, or slow learning in just a few sessions. Likewise, supervisors may expect you to handle "just one more client" and assure you, "It is very much appreciated, you'll see. I'll make this up to you. Just help me out of this jam."

COMMON SYMPTOMS, STANDARD SOLUTIONS

A little stress is not a bad thing, because it can sharpen your focus and give you an adrenaline boost in an emergency situation. Prolonged work-related stress, however, can cause headaches, chest pain, shortness of breath, stomachaches, tiredness, and sleep problems. This may in turn cause anxiety, irritability, mood swings, resentment, and burnout. The effects on behavior can include not eating, overeating, having angry outbursts, crying, and experiencing decreased productivity.

Traditional solutions to life stressors include exercising, relaxing, getting lots of sleep, and in some cases getting counseling or psychotherapy.* These standard stress-management approaches do not deal with the *cause* of stress, however, and a contrasting approach is to promote *organizational change* as a solution.† This involves analyzing the stressful factors in a work environment such as excessive workload or conflicting expectations and making changes in your environment to reduce the stress. In addition, you will want to consider the Big Three factors that if properly balanced can mitigate chronic stress: proper diet, adequate sleep, and sufficient exercise. It can easily happen that you are working so much and so intensely that you are not taking care of your physical health. Living on fried, quick, greasy fast food; getting only 5 hours

> "The foundation of stress prevention is built on a healthy diet, 8 hours of sleep, and vigorous exercise at least three times per week."

---

* See the Mayo Clinic Web site: mayoclinic.com/health/stress-symptoms.
† See cdc.gov/niosh/stresswk.html

of sleep each night; and engaging in virtually no exercise are a prescription for an emotional or physical breakdown. The foundation of stress prevention is built on a healthy diet, 8 hours of sleep, and vigorous exercise at least three times per week. If you don't have this regime in place, consider improving your diet, sleep, and exercise habits to help you deal with other stressors in your life.

ORGANIZATIONAL CHANGE TO HANDLE STRESS

From a behavioral perspective, the organizational change approach makes much more sense than taking up "choiceless awareness meditation," listening to nature sound CDs as you drive down the road, or taking hatha yoga classes on Thursday night. If you are overwhelmed by your caseload, find that you are inhaling Reese's Pieces by the handful, and are constantly on the lookout for another "Happy Hour 4:00 till 8:00" sign, taking a Pilates class is not the right solution for you. Pilates and other forms of exercise will make you strong and fit, but exercise alone will not fix your stress. You have to deal with the actual problem. How did you come to have this caseload anyway? Was it a failure to say "no" at just the right time? Are you having difficulty closing out cases? Do you need some help with a difficult case but don't know whom to talk to?

So what can you do to better manage your life? For starters, we recommend opening with a review of the tips in Chapter 17 on time management. Some of your stress may result from not having a reasonable daily and weekly plan that allows you to efficiently get from one consulting site to the next. If you become stressed as a result of dealing with traffic congestion, look at your schedule to determine if there is a more efficient way to get from A to B to C. Perhaps some of your clients would be willing to change their standing appointments; this might allow you to avoid the morning rush. If some of your stress is related to your always being late, take a close look at factors that could be causing the lateness. One colleague had this frustration and found that he had to take the blame for not informing his clients that he had to be out the door at a fixed time; he left it vague, thinking that it was rude to cut off a conversation in mid-sentence. By informing these clients at the beginning of the consult of his required departure time, he was able to prompt them with, "OK, Mrs. Rudd. I have only 5 minutes before I have to leave." By exercising a combination of assertiveness, time management, and personal communications skills, he was able to leave on time and avoid speeding through yellow lights to get to his next appointment.

Further analysis of your workweek may reveal other opportunities for saving time. These ways include employing a more efficient method of report writing, using software on your PDA to keep track of billable hours, or training mediators with video clips instead of role-plays. You might discover that it is more efficient to return phone calls and e-mail in batches rather than at odd times throughout the day. You might want to consider handling these daily chores using the Premack Principle (as described in Chapter 17), that is, if you have completed all your written work for the day, you can spend an allotted amount of time on phone calls so that calls don't add up all day long.

EXTERNAL STRESSORS

Perhaps one of the most delicate, but important, ways to reduce stress is to analyze the nature of the work you are doing to determine if it is part of the cause. There was probably a point early in your present job where you felt that you had everything under control. You understood your client's problem, you had cooperation from your mediators, your behavior

> "Perhaps one of the most delicate, but important, ways to reduce stress is to analyze the nature of the work you are doing to determine if that is part of the cause."

program was getting good results, and you got recognition and a bonus from the CEO. Life was good. That might have been a year or two ago, and since then somehow everything got out of control. What exactly happened? Did you agree to supervise two first-year students, only to discover that they required a lot of hand-holding? Or was it the case of the single mother with the autistic child who was progressing so nicely and then somehow hit a plateau? This child's mother started to ask questions about applied behavior analysis. She was reading articles about special diets and wondering if maybe applied behavior analysis wasn't the solution after all. Picky, picky, picky. It was nerve-racking; she just wouldn't be satisfied with what you told her and wanted to talk to your supervisor. And then your supervisor took the mom's side and started questioning your program plan. Now *that* can make a person jumpy.

And remember when your roommate suddenly decided to leave town, and you got stuck with the whole month's rent until the lease ran out? This is when you started wearing the T-shirt that said, "STRESS: The confusion created when one's mind overrides the body's basic desire to choke the living daylights out of someone who desperately deserves it." Thanks to your roommate, you had to take on three extra cases to make up the difference in rent money: "I'm young. I don't have a boyfriend right now, and I can do 50 hours a week, at least until I find a roommate."

WHAT TO DO ABOUT STRESS: A BEHAVIORAL APPROACH

Much of the traditional literature and advice on the Internet has to do with identifying the signs of stress and taking steps to manage the symptoms. Behavior analysts who trust their ability to understand human behavior, including their own, are likely to reject this strategy in favor of a more comfortable, familiar behavioral approach. After all, if a client came to you with all of these symptoms, wouldn't you start with pinpointing the behaviors and doing a functional analysis?

It makes sense, of course, to have the big picture on stress. Stress can come from major life changes such as recently getting married (or divorced), getting pregnant (especially if it is an unplanned pregnancy), moving to a new city, or changing jobs. A new neighbor with a barking dog that interrupts your sleep, a roommate who keeps odd hours, or the unemployed guy in the upstairs apartment who apparently got a drum set for Christmas and is experimenting with his flams and hi-hats all night long will certainly cause stress for those in close proximity. Personal family issues can also throw a person off: a spouse who suddenly decides to quit work, in-laws who are meddling, or an old flame who finds you on Facebook and wants to get back together.

Step 1: Pinpoint the Stressful Emotions, Feelings, and Behaviors

As with any situation you would encounter at work, you know the first thing you must do is pinpoint the behavior or result that is causing the problem. Is it actually physical? Chest pains, shortness of breath, anxiety? If so, a medical checkup by your family physician is certainly called for. What about behaviors that you've adopted that look like they could cause trouble down the line: overeating, crying spells, or decreased productivity? You know how to make lists and data sheets, so start with this. Chart your symptoms (both behavioral and physical) to determine the frequency, and make a column for possible antecedents. You should note the time of day as well as coincidental events. For example, if your anxiety goes up around 3:00 p.m. on some days but not others, look at your appointment book or iCal and see what was going on leading up to 3:00. Is it racing through traffic or dealing with your family that appears to be sabotaging your own healthy living treatment plan? Does this happen only on days when you were up late the night before and didn't get enough sleep? You can analyze your data by using a standard ABC analysis (antecedent–behavior–consequence) with the addition of "Setting Event" as a broad category of rather vague issues that might set the occasion

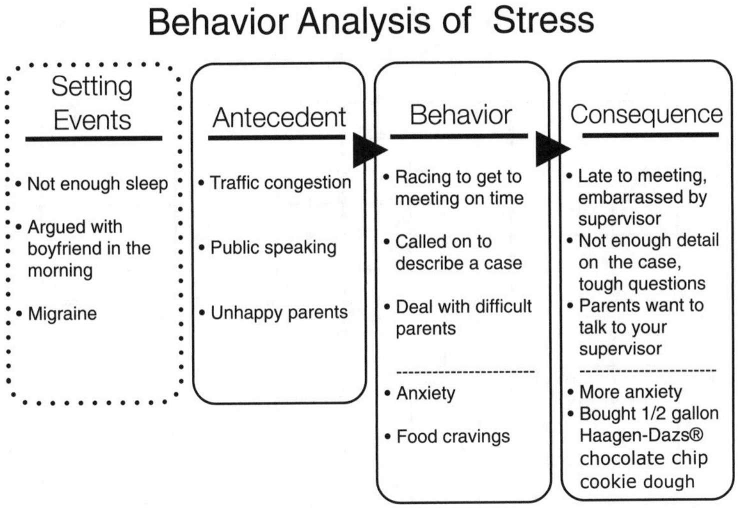

Behavior Analysis of Stress

Figure 19.1 A behavior analysis of stress.

for a range of unpleasant experiences during the following day (see Figure 19.1).

Step 2: Perform a Functional Analysis

This is going to be tricky because you are essentially experimenting with yourself. Be objective about this, starting with recording instances of problematic behavior and troublesome emotional states. A scatter plot graph that you fill in each time you experience anxiety, chest pains, or food cravings might look like the one in Figure 19.2. These are color coded to show black for chest pains, dark gray for anxiety, and lightgray for food cravings. Next, it is necessary to look back through your daily planner to identify which events were associated with these symptoms of stress in your life.

To determine if your hypothesis is correct and you have in fact identified the controlling variable, you will need to make a temporary change in your schedule or sequence of events. Consulting on three intense client cases back-to-back without a break may be just too much, or racing across town for a supervision meeting

Scatter Plot of Times of Day that Occasion Emotional Reactions

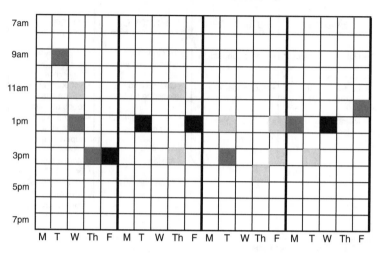

Figure 19.2 Scatter plot graph of times of day that cause an emotional reaction. *Note.* Black is for chest pains, dark gray is for anxiety, and light gray is for food cravings.

with no time to review your notes may be too much. By changing your schedule and honestly and objectively recording the results, you should be able to isolate the factors that are causing stress in your life. Some variables will be difficult to manipulate, especially if they occur infrequently, such as too little sleep or an argument with your boyfriend. Are you staying up late because you drank too much coffee at dinner or you have a habit of watching *The Daily Show* and *The Colbert Report* every night or because you possibly have a disturbed sleep pattern (sleep apnea)? If the problem is the latter, your physician should be able to run the necessary tests to isolate the medical condition. If your sleep problems are related to your getting wound up when you watch late-night television, it may be time for you to invest in a recording device

"You owe it to yourself to apply what you know to solve your own personal problems."

so you can watch your favorite nighttime shows at an earlier time. This is the sort of functional analysis you would perform when a client expressed concern about anxiety or poor performance. You owe it to yourself to apply what you know to solve your own personal problems.

Step 3: Develop a Short-Term Intervention

Depending on what you discover as a result of your functional analysis, it may be possible to work out a short-term solution that will provide some relief while you work on a long-term solution. For example, if you discover that your stress level goes down during the day when you get more sleep the night before, a quick fix might just be a matter of better time management and some discipline with regard to television shows or late-night phone conversations with friends and relatives. The same goes for variables that produce stress during the day. Being more firm with your supervisor and making it clear that when you finish these three overload cases, you will not pick up any more could go a long way toward relieving your stressful week. Likewise, when working with a client who appears to be dissatisfied with your services and is about to complain to your boss, consult with your supervisor before the client does (you'll feel tremendous relief). Your supervisor might provide the resources to help you with the client or decide that it is time to transfer this client to another behavioral consultant.

Step 4: Develop and Adopt a Long-Term Plan

Your long-term plan, which you might be working on to bring to fruition in 6 to 9 months, could involve some rather major changes in your life. Getting rid of the argumentative boyfriend or girl-friend, changing jobs, moving to a less crowded part of the state where the living is less hectic, setting up your own private practice where you work for yourself, and moving out of a noisy apartment into your own home are all examples of ways to relieve stress that involve a good deal of planning and possibly some hardship and stress in the short term.

SUMMARY

Being a behavior analyst can be both a blessing and a curse. It is truly a blessing to be a professional who possesses a thorough and deep understanding of human behavior and the skills necessary to change the lives of others in a positive manner. The curse is that everyone around you expects you to be able to solve your own behavior problems. This is not entirely incorrect; in our field we do prepare students with the motto, "Behavior analysis begins at home." You *should* be your own best behavior therapist, you should be aware of the effect that your environment has on your behavior, and you should know when something is not right with your health or your stress level.

In addition, you should be able to analyze your own behavior and provide your own interventions. Begin with making sure you have your physical and emotional health in order with proper diet, sufficient sleep, and vigorous exercise. Then conduct a functional analysis, and develop a short-term intervention. You've had plenty of practice using your skills with clients. Now it's time to use them to improve and enhance your own life.

20
Knowing When to Seek Help
(and How to Receive Feedback)

Feedback is the breakfast of champions.

Ken Blanchard

"Thanks for coming in today, Michael. Unfortunately, what we need to talk about is not so pleasant. The facility is very unhappy with your work and wants another consultant."
"You're kidding! I thought I was doing a really good job."

When we interviewed owners of behavioral treatment clinics and senior behavior analysts charged with training new Board Certified Behavior Analysts®, they expressed the same concern over and over. Their new hires, although young, eager, and full of enthusiasm, were often totally unaware of their professional shortcomings. As a result, they would get deep into a case, miss the early warning signs of imminent trouble, and begin to flounder. The interviewed supervisors recounted that in most cases it took a phone call from a disgruntled client to alert them to the situation. In nearly every case, the young enthusiastic behavior analyst did not know there was a problem. In the few cases where the beginner knew there was a problem, he or she didn't seek help, for whatever reason.

There are always warning signs when things are not going well, but you have to be sensitive, very sensitive, to what is happening around you to catch the signs that someone is in trouble (and it might be you). In many ways, the warning signs are the same cues that show up in interpersonal relationships that are headed south. Phone calls or e-mail are not returned, meetings are cancelled, or key people don't show up to meetings. Direct contacts are not warm, eye contact is brief, and there is little informal conversation at the beginning of a meeting. An experienced consultant will know there is something wrong the first time she sees this shift in a relationship. New consultants will often miss the red flags or dismiss them, thinking, "I guess he was just having a bad day." We don't necessarily train behavior analysts in the subtleties of body language, although we should. We might be spending too much time teaching behavioral consultants about autoclitics and intra-verbals (Skinner, 1957) when we should be focusing more on reading facial cues and body language (Harvard Business School Press, 2004a; Klaus, 2007). When you see a frowning administrator push himself back from the table in a performance management setting, you should know if you don't do something quick, there is a good chance your contract will not be approved. Reading body language and subtle signs is a skill that is difficult to teach in the classroom. Unfortunately, this skill often has to be learned the hard way, through experience. You can speed that process along if you are aware of a few common signals that things are not going well.

SOCIAL CONTINGENCIES OF SUPERVISION

Some unspoken contingencies seem to drive a situation where things are going wrong. To admit early on that you don't know what you are doing is humiliating, and the hope that this will somehow all work out, or is just a bad dream, prevails. Beginners are afraid that asking for help is a clear sign of weakness. To admit you are in over your head is to confess ignorance of a very basic sort. People get fired for that, and if the word gets around, you might not get hired anywhere else. This is a very scary thought. If you indicate you think there is a problem and there isn't, you've wasted your supervisor's time. Maybe you are expected to solve these problems on your own and not bother people. And on and on goes the thinking that prevents you from getting the help that is needed.

What new behavioral consultants and behavior analysts don't seem to understand is that the organization wants very much for them to succeed. In most cases, supervisors and managers will go to great lengths to provide the support, supervision, and extra training necessary for you to thrive. They would rather you come in with no deficits, of course, but good people are hard to find. If it takes a few extra hours of supervision in the first 6 months to make sure you master your craft, then it will be provided. Otherwise the company looks bad, and it certainly doesn't need negative PR with regard to its employees. Furthermore, there are high costs associated with replacing professional staff.

WHEN TO SEEK HELP

The ideal consultation runs smoothly from start to finish. Your initial meeting with the client is cordial, you work hard to make a good first impression (Bixler & Dugan, 2001), and you are well received and immediately begin using your Dale Carnegie skills of listening and reflecting. You gain access to the setting or spend time with the child or adult client, complete your analysis, and develop a behavior intervention plan. You pitch the plan, negotiate the details, obtain buy in, select the mediator, begin training, and watch the magic happen.

Here is where things can go wrong. The mediator might not be comfortable in her new role as trainer or contingency manager. Perhaps something went awry the first time she gave out a reinforcer or applied a consequence. Suddenly she's not so sure about this behavioral approach. But she's afraid to offend you, because you've put so much work into this project, and you are such a nice, enthusiastic person. The manager says nothing directly, but she is quiet during the weekly meeting. She responds to "How is it going?" with a short "Fine." If you take this reply literally, you will be making a big mistake. "Fine" means "I'm uncomfortable with this procedure, and I'm not sure I can do this consistently." If you watch her body language, you'll see that she is looking at her shoes most of the time and fidgeting with a pen, her answers are short, she squirms in her chair, and she wants this meeting over. Now, if you were being direct, you would say, "I get the sense that you don't like what we're doing here. What is your problem?" But, of course, you can't be direct or you'll lose your mediator and perhaps the whole project.

Anish was working totally on his own with his first business client. He had been shadowing a senior consultant for 6 months and was ready for his big test. The initial meetings at the insurance company went well, and he had just completed his first round of stand-up training with middle management. Over 3 days he covered the basic principles of performance management, and now he was meeting with department heads on their turf. Greg Shephard, the director of sales, was in his early 50s. Anish remembered Shephard as the guy in the back who kept his arms folded and looked down at his notebook during most of the workshop. Anish did his best to start off easy with some casual, light conversation about the weather and the upcoming bowl game. He then gave Shephard a chance to describe the pinpoint he wanted to address first in his department. "Can't think of anything right off," Shephard said. "OK, let's start with this spreadsheet that shows your field supervisors are not reporting on their contacts and providing leads for the call center," Anish offered. "Well, OK," was the response that Shephard gave, with flat affect and no eye contact. Anish struggled with the sales department head for the next 2 days and finally got a written plan of action. In an e-mail report to his supervisor, Anish described his progress but didn't mention Shephard's foot-dragging. Anish dismissed the lack of enthusiasm as just part of the game that clients play where they don't act happy, reluctantly agree, and just hope these consulting guys will go away. They always do. Anish's meetings with the other department heads over the next month matched his first encounter. Anish wasn't sure where he stood exactly. He couldn't say that any data were coming in from the company, but he thought that was typical. And then he got the call. It seems the CEO of the insurance company called his CEO and said, "It's just not working out. We'd like to cancel the rest of the contract as per our written agreement on page 4." No other explanation was offered, and Anish was called in to headquarters for a complete debriefing. It was humiliating to have to go through the details, but now he could see the signals were there all along: cancelled meetings, deadlines missed by department heads, and e-mail not returned. The cool responses to lunch invitations all suddenly looked like flashing neon signs on a cold dark night.

SOCIAL CUES FOR BEHAVIORAL CONSULTANTS

Most of applied behavior analysis is an intense social enterprise. As a professional consultant, you have to sell yourself first and then sell your product. Your product is our

> "Most of applied behavior analysis is an intense social enterprise."

evidence-based approach to performance improvement. If you are not well liked, this is not going to work. If your client doesn't trust you, this isn't going to work. If your customer service isn't better than the competition's, you will absolutely lose the client. A common mistake of new consultants is they focus too much on their presentation and not enough on how the client is receiving and responding to the information. You certainly want to watch for the early signs that you've hit a rough spot; a client who looks confused (look for facial signs: a furrowed brow, squinting eyes, pursed lips) is a signal for you to stop and say something like, "I can see that I haven't explained this very well. Can I give you a better example?"

Get the client talking, listen attentively, and try to get this back on track. The benchmark for success is a relaxed, smiling, head-nodding client who feels comfortable interrupting you to give his own examples and share an anecdote. A client who is staring past you, fidgeting in her chair with her shoulders hunched forward, and tapping her fingers is a sign that things are not going well. This person is not tracking your argument and does not see the relevance of what you are talking about. If she put it into words, she would say, "I don't get this, and I don't think it applies to me. This is not the way I would approach our performance problems. I wish you would go away." Of course, there are differences from one person to the next, and cues may vary. For some people, a far-away look means they are disinterested; for others, it could mean they are concentrating. The basic message is to watch for any signs that your client is bored, disinterested, confused, unsure, anxious, or worried. If you do not intervene immediately, your client will move to the next stage: frustrated, fed up, and ready to disengage. This is when she will ask the CEO to make the phone call to tell you it has been nice working with you but …

WHAT SHOULD YOU DO?

At the first sign your client is the least bit unhappy with your service, contact your supervisor, talk on the phone, and ask for a meeting ASAP to discuss the situation. Describe the circumstances from the beginning. Don't leave out embarrassing details, be objective, and don't apologize. Make a sincere statement that you want to learn and improve and that you are anxious to get some feedback and perhaps some additional training. Make clear your commitment to the company or agency.

RECEIVING FEEDBACK

Step 1: Request Feedback

If this is your first experience, you may have some trepidation, but put yourself in the mind-set of, "This is going to be good for me.

I'm ready to learn, I'm working with experienced people, I'm in good hands, and I'm ready to grow as a behavioral consultant."

Step 2: Take Notes

Your feedback session could take 30 minutes to an hour. Many points will be covered, and you will not be able to remember everything. Plus, you'll want to impress your supervisor and show that this is important to you, that it will make a lasting impression.

Step 3: Listen Intently, Ask Questions, and Be Engaged

This is not a tongue-lashing where you are expected to sit there and take your punishment. This is your supervisor sharing years of experience, and you need to listen to every word and follow her arguments closely, point by point. Your body language needs to show that you care about what she says, that you get it, and that you understand. Don't interrupt. Wait until she gives you a cue to respond: "So, Jamie, any questions so far?" This is where you look down at your notepad and present your first question. Do not respond, "No, I guess not." Act friendly and professional, don't cry, and don't dissolve into nervous laughter or giggling. And offer no "buts" either. If you act defensively and try to explain away your behavior by saying, "But, I just thought … ," you create the impression that you cannot believe you made a mistake. Your supervisor is not looking for you to prostrate yourself or apologize profusely. She wants you to show that you've been paying attention, that you care, and that you are eager to learn from your mistake. A really good question should give your supervisor reason to launch into another paragraph or two of explanation. Continue to take notes, and be

> **"Your supervisor is not looking for you to prostrate yourself or apologize profusely. She wants you to show that you've been paying attention, that you care, and that you are eager to learn from your mistake."**

ready for your next opportunity to engage in the dialogue. Don't argue, take offense, or try to justify the actions that brought you to this point. It won't do any good, and it is likely to irritate your supervisor. Watch the supervisor's body language. You should be able to tell when she is about to finish the meeting. At the end, quickly run through the checklist that you made so the supervisor knows you got all the key points. Shake hands, smile, tell your supervisor how much you appreciate her time, and indicate you will follow up with an e-mail, thus putting the key points of this meeting in writing. Ask when you should schedule a follow-up meeting.

Step 4: Make a Record of Your Feedback Session

As soon as possible, create a written document of the feedback session based on your notes. Send it to your supervisor as an attachment to an e-mail. Again, thank the supervisor for her time. Put your notes in bullet points with headers as appropriate so it is easy for your supervisor to read. If you have analyzed the feedback and can see that things that you should change are time based, indicate this in the document; for example, "Repair relationship with teacher aide; start on Nov. 10" or "Ask for meeting with OT at next habilitation team meeting."

Step 5: Develop a Plan of Action

Fixing whatever went wrong is going to take several action steps over the next several weeks. Don't lose track of these steps. Put them in your calendar and also on a to-do list. Clearly indicate what you plan to do by listing phone calls to be made and e-mail to be sent. These steps should be defined in terms of both process and outcome.

Step 6: Report Back

Send an e-mail to your supervisor every 2 or 3 weeks to report on your progress. These messages don't need to be lengthy memos. Short notes indicating you had a meeting, turned in an evaluation, or had lunch with the principal are enough to let your supervisor know that the time she spent with you was worthwhile, that you took it all seriously, and that you are taking corrective steps.

Step 7: Ask for a Follow-up Meeting

A month after the first meeting (or sooner if the supervisor requests it), it will be time for a second face-to-face meeting with your supervisor. This is your chance to once again thank her for her time and advice and to describe the steps you've taken and the status of the situation. If you have done everything right, this meeting should be fairly short, and you'll do most of the talking. You may have a couple of questions, but, we hope, you will have solved most of the problems and will be back in good standing with your company. Your supervisor should be pleased that you took the feedback seriously and acted promptly to take corrective steps. A firm handshake, a big smile, and a final very sincere "thank you for everything" should conclude this series of interactions.

SUMMARY

Being an effective behavioral consultant is a complex, multifaceted job that involves not only theoretical and technical knowledge of human behavior but a huge dose of human relations training as well. Knowing how to make a great presentation and how to maintain active, positive engagement with your client for months at a time is an essential consulting behavior. The more experience you have, the better you will be at this. Pay attention. Try hard. Be prepared to learn from every encounter. And realize you cannot make everyone happy. It would be nice, but all of your clients do not have to love you. You should, however, conduct yourself in a way so your clients respect you. Earning respect for a job well done will provide great satisfaction for the consulting behavior analyst.

FOR FURTHER READING

Bixler, S., & Dugan, L. S. (2001). *5 steps to professional presence.* Avon, MA: Adams Media.

Harvard Business School Press. (2004a). *Face-to-face communications for clarity and impact.* Boston: Author.

Harvard Business School Press. (2004b). *Manager's toolkit: The 13 skills managers need to succeed.* Boston: Author.

Klaus, P. (2007). *The hard truth about soft skills: Workplace lessons smart people wish they'd learned sooner.* New York: HarperCollins.

Skinner, B. F. (1957). *Verbal behavior.* Englewood Cliffs, NJ: Prentice Hall.

Section

Five

Advanced Consulting Strategies

21
Critical Thinking

What skeptical thinking boils down to is the means to construct, and to understand, a reasoned argument and—especially important—to recognize a fallacious or fraudulent argument.

Carl Sagan

Behavior analysts are by nature critical thinkers, skeptics even. We are not pessimists or optimists. Rather we often adhere so close to the "Where's the data?" line that many people in human services find us to be somewhat strange. We always question everything and everybody, even those with dramatic personal anecdotes—especially those with dramatic personal anecdotes! Because we won't accept on face value a touching story of a new miraculous breakthrough cure for a behavior problem, to some it might appear that we are just basically cynics who don't believe in anything but science.

Some of this is true. We *do* require everyone who wants to talk about behavior to have data, and not just any data but data that require repeated measures, observational reliability, and, of course, social validity. And don't forget, there must be a clear demonstration of experimental control. And did we mention that the data have to show a socially significant effect? They can't be just cooked-data randomized statistical test mumbo jumbo. By this standard, 98% of the evidence about human behavior in the world falls short. No other treatment approach comes close to this rigor, and it can make colleagues in related professions, in particular, and quite a few consumers squirm. We are not interested in general correlational findings (even if they are statistically significant) that show that boys are better at math or that people who sleep on their backs are repressed (or is it depressed?). We *are* interested in knowing the causal variables surrounding human behavior, and we understand that some are proximal (close in) and some are distal (somewhat removed, such as something that happened 2 months ago). Because we can't control the distal variables, we emphasize the proximal and insist that to *really* understand behavior, we must systematically make manipulations, interventions, treatments, and any changes in the environment. We believe in a treatment only when a sufficient number of replications have been performed. Even then, we understand that these findings, perhaps published by reputable applied scientists in our flagship journal, may not directly apply to our current client. In these cases, we insist on taking a baseline for each client behavior, testing the intervention, and determining for ourselves if it works in *this* setting for *this* individual. As it turns out, we are intermittently reinforced for critical thinking, as it frequently happens that treatments that sound too good to be true aren't true. And it happens every day that we try a published procedure with a client and discover, for some reason, it doesn't work. We have mixed emotions over this. Although we wished that the treatment had worked and were disappointed that it didn't, we were certainly glad our baseline and subsequent data collection objectively evaluated the treatment.

CASUAL THINKING VERSUS CRITICAL THINKING

Most thinking that we do on a daily basis is not critical thinking but *casual thinking* or problem solving: "It's time for an oil change. I pass a Jiffy Lube on my way home from work every afternoon, so I guess I'll stop and have it done today." Other routine thinking may involve taking advice from a friend: "I want you to try this Fair Trade Organic Sumatran Reserve coffee. It has the most amazing flavor and is grown in soil that has never been treated with pesticides, so you won't get cancer from it." It takes some effort to do critical thinking on matters such as this, and depending on whether you share your critical thinking with others, you might not be too popular. People like to influence one another, and if you analyze other peoples' recommendations and critique their advice, there is a good chance they will seek out other company. In most cases, casual thinking is perfectly fine. You can get through the day without any serious damage to your friendships, although you might buy an overpriced bag of coffee that tastes exactly like Folgers. The real problem occurs when you allow yourself to be lulled into mushy, casual thinking when valuable time, money, and opportunity costs are at stake.

> "The real problem occurs when you allow yourself to be lulled into mushy, casual thinking when valuable time, money, and opportunity costs are at stake."

CRITICAL THINKING IN ACTION

Our version of critical thinking, having evolved directly from our single-subject experimental method and honed over four decades in applied settings and university laboratories, has left us with a legacy of skepticism about nonbehavioral theories and treatments. We run into theories about behavior daily in our role as professionals, and new theories are being announced nearly every day. From chiropractic manipulations to improve the nervous

> "We run into theories about behavior daily in our role as professionals, and new theories are being announced nearly every day."

system, to chelation administrations to remove heavy metals from the body, to gluten- and casein-free diets that supposedly reduce peptide levels and improve behavior and cognitive functioning, new untested theories abound. Using casual thinking in these cases could result in danger to a client and a major hit on the bank account, not to mention a large amount of time wasted. Families desperate for any shred of hope for a cure for autism seem disinclined to engage in critical thinking about treatment effectiveness.

We can look at one very popular treatment, sensory integration (SI), as a way of showing what would be involved in casually and critically thinking about an untested theory. SI, a wildly popular theory, was introduced in 1972 (Ayres, 1972). The claim is that the integration of stimuli from the body and the environment requires a "balance between excitatory and inhibitory neurological systems" (Bundy & Murray, 2002). Sensory integrative therapy (SIT) includes doing activities to stimulate the vestibular system, such as being pushed in a swing or rolled on a mat or riding on scooter boards (Smith, Mruzek, & Mozingo, 2005, pp. 331–332). Other activities such as squeezing the client between gym pads to provide "deep pressure" or brushing him with a soft hairbrush are postulated to therapeutically stimulate the proprioceptive or tactile systems of the individual. As far-fetched as this therapy sounds, supporters of SIT, usually occupational therapists, claim that these treatments enhance the individual's ability to focus on materials, reduce his maladaptive behavior, and lead to improvements in nervous system functioning (Smith et al., 2005, p. 332).

Casual thinking might go like this: "The theory has been around a long time, there is research on it, and occupational therapists recommend it. Why not try it?"

Critical thinking (Paul & Elder, 2002) requires people to separate *information* (data) from their *assumptions* about the information. We need to be able to see a straight line from *data* to *inferences* (conclusions), recognizing our assumptions and reducing them to a minimum. Finally, we have to understand that the

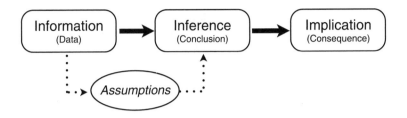

Figure 21.1 This flowchart shows critical thinking steps, including the role that assumptions can play in the critical thinking process. Ideally we are fully aware of our assumptions and can discount them to the fullest extent possible so that the inferences we make are based solely on the data that we have at hand. Adapted with permission from Paul, R. W. and Elder, L. (2002). *Critical thinking: Tools for taking charge of your professional and personal life.* Upper Saddle River, NJ: Pearson Education.

inferences we make have *implications* (consequences), in this case the implications for the client (see Figure 21.1). To accept this theory without critically investigating the published research and without evaluating it for individual clients is a clear example of noncritical thinking that can lead to wasted time, squandered resources, and false hope.

Therapists who employ SIT are working without information that shows under laboratory conditions that the proposed procedures are effective, and they are assuming the theory is correct. Unfortunately, this leads to their spending hours and hours engaged in wasteful and repetitive brushing, swinging, and squeezing of their clients to no avail. Consumers of these services are clearly not engaging in critical thinking, because if they were, they would be asking the occupational therapist, "Is that true? Can you give me more details on that? Does this really make sense?" Consumers cannot be expected to be knowledgeable about research methodology or theories involving the nervous system. They are at the mercy of these professionals to properly characterize the treatment approach. There is a certain amount of sophistry going on here as well; supporters of SIT should be well aware that their research is weak and inconclusive.* In a detailed analysis of this body of work, Smith et al.

---

* *Sophistry* is a specious argument used for deceiving someone.

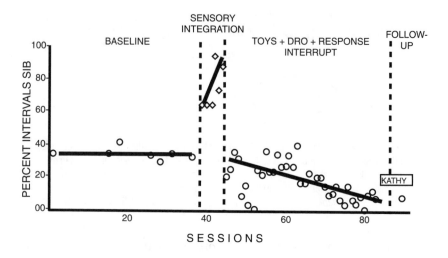

Figure 21.2 Data for one participant from the Mason and Iwata (1990) study. Data are redrawn to more clearly illustrate the effects of the sensory integration procedures. The sensory integration techniques used were designed by occupational therapists who used them regularly. From Mason, S. A. and Iwata, B. A. (1990). Artifactual effects of sensory-integrative therapy on self-injurious behavior. *Journal of Applied Behavior Analysis, 23,* 361–370.

(2005, p. 345) concluded, "Studies indicate that SIT is ineffective and that its theoretical underpinnings and assessment practices are unvalidated."

One behavioral study done with sensory integration (SI) as an independent variable (Mason & Iwata, 1990) was carried out with the specific purpose of testing SI theory. Results showed that none of the three participants improved with SI treatment. The data for one participant are shown in Figure 21.2 for illustrative purposes.* Kathy clearly showed a paradoxical effect, that is, she showed increased self-injurious behaviors when the SI procedures were put in place, which certainly was not an expected outcome but one that indicates that such essentially untested, but great-sounding, methods can in fact do some harm.

---

* The data have been redrawn to isolate the effects for one client, Kathy. In the original figure, hers is the middle graph of three in a multiple baseline design.

FACILITATED COMMUNICATION: THE POSTER CHILD FOR FAILED CRITICAL THINKING

Perhaps the greatest debacle tied to a failure of critical thinking involved facilitated communication (FC), a fad treatment if ever there was one. FC began in Australia as a treatment for individuals with cerebral palsy and then migrated to the United States in the early 1990s as a method for assisting communication for people with autism (Jacobson, Foxx, & Mulick, 2005). The assumption is that individuals with autism have "undisclosed literacy" and that a facilitator can assist them in bringing out their latent literary talents by helping them type on a keyboard to express themselves. It wouldn't take much critical thinking to bring this observation (and assumption) to its knees, so it is surprising that FC caught on and spread like wildfire in the late 1990s. So desperate and gullible were the consumers (i.e., family members, school administrators) that facilitators were hired at school system expense to sit with severely and profoundly mentally handicapped individuals and help them write poetry, express lifestyle preferences, and, in several cases, allege crimes against relatives. Even the judicial system failed in its critical thinking skills to see the obvious: It was the facilitators who were writing the short stories and doing the math problems. One dead giveaway was that the individuals who were being facilitated did not appear in the least to be interested in the task and often were observed to be looking in the opposite direction of the keyboard (Foxx, 1994). Instead of asking critical questions about the procedures, advocates for FC focused on the rights of individuals with disabilities and *assumed* that they were capable of self-advocacy, thus totally distorting the information in front of them. The consequences (see Figure 21.1) have been devastating for families that were ripped apart by fallacious allegations. Lives were ruined when innocent people spent months in jail until a proper hearing and testimony from experts could be provided (Maurice, Green, & Luce, 1996).

CRITICAL THINKING GOES TO WORK

For you as a behavior analyst, critical thinking tools are essential every time you go to work. Teachers, parents, and administrators will describe some horrendous situation that requires your immediate

> "For you as a behavior analyst, critical thinking tools are essential every time you go to work."

attention and will confront you at least weekly: "Drop everything, and take care of this!" It is usually a thirdhand story about some individual who is in desperate need of "behavior modification." Your job is to remain calm, review the evidence, determine what assumptions have been made and by whom, and do your level best to arrive at a reasoned conclusion. People on the front lines have been reinforced for exaggerating their claims and embellishing their stories. Anything less will get a shrug and a smile and little else. Most people do not have data to give you to quantify the problem, and colorful anecdotes full of dramatic details designed to spur you to action are the modus operandi. You must be especially careful not to believe the first story you hear but rather to reserve your judgment until you've heard from all parties. Then, in full critical thinking mode, try to sort out what happened and arrive at a conclusion. If you want to remain true to your behavior analysis tradition, you will need to establish some sort of actual baseline before you draw any conclusions. Establishing this baseline is key to critical thinking in applied behavior analysis and will no doubt cause consternation among those who want immediate action.

The second challenge to your critical thinking will come when you devise an intervention based on your functional analysis and implement a treatment. Because it is your treatment plan, you will be inclined to like it and believe it will work. You will need to put on your critical thinking hat here and evaluate your plan objectively, setting aside your assumption that "Of course it will work, I designed it. Why wouldn't it work?" The correction against going easy on yourself is to submit your program and data to a peer-review committee and receive feedback on a regular basis.

Critical thinking comes into play every time you open a journal to research the current best practices in behavior analysis. Although we usually think of "published in a peer-reviewed journal" as the standard for evidence-based practice, experience has shown that a significant number of such studies fall short, way short. Applying critical thinking to

> "Critical thinking comes into play every time you open a journal to research the current best practices in behavior analysis."

these studies reveals that many simply don't meet our standards. Baselines are too short or too variable, the dependent variable is not well defined, the reliability is below 80%, conditions are not replicated, the size of effect is too small to be socially significant, and more. Research conducted in the first author's Behavior Analysis Research Lab (Normand & Bailey, 2006) showed that participants, all Board Certified Behavior Analysts®, made accurate decisions for only 72% of the graphs they reviewed. Even the addition of celeration lines did not improve overall accuracy. If BCBAs (presumably well trained at the master's level) are unable to properly analyze published studies and determine which are suitable as a basis for building an effective treatment plan, it is clear that we have a critical thinking problem in our profession.

SUMMARY

Finally, we should point out that opportunities for critical thinking arise each and every day in professional practice. Colleagues from other professions will advocate the use of their favorite interventions, parents will want you to approve the use of a strange new fad they read about on the Internet, and administrators will push you to approve far fewer hours of treatment than has been shown to be effective in the quality literature. In all these cases, you will need to combine your critical thinking skills with several other of the 25 skills discussed in this book: assertiveness, ethics in daily life, persuasion, lobbying, handling difficult people, and problem solving.

FOR FURTHER READING

Critical thinking is a skill that can be learned, and there are several good books available for those who want to brush up or catch up.

Ayres, A. J. (1972). *Sensory integration and learning disorders.* Los Angeles: Western Psychological Services.

Bundy, A. C., & Murray, E. A. (2002). Assessing sensory integrative dysfunction. In A. C. Bundy, S. J. Lane, & A. Murray (Eds.), *Sensory integration: Theory and practice* (2nd ed., pp. 3–34). Philadelphia: Davis.

Foxx, R. M. (1994, Fall). Facilitated communication in Pennsylvania: Scientifically invalid but politically correct. *Dimensions,* 1–9.

Jacobson, J. W., Foxx, R. M., & Mulick, J. A. (Eds.). (2005). *Controversial therapies for developmental disabilities.* Mahwah, NJ: Lawrence Erlbaum Associates.

Mason, S. A., & Iwata, B. A. (1990). Artifactual effects of sensory-integrative therapy on self-injurious behavior. *Journal of Applied Behavior Analysis, 23,* 361–370.

Normand, M. T., & Bailey, J. S. (2006). The effects of celeration lines on accurate data analysis. *Behavior Modification, 30,* 295–314.

Paul, R. W., & Elder, L. (2002). *Critical thinking: Tools for taking charge of your professional and personal life.* Upper Saddle River, NJ: Pearson Education.

Smith, T., Mruzek, D. W., & Mozingo, D. (2005). Sensory integrative therapy. In J. W. Jacobson, R. M. Foxx, & J. A. Mulick (Eds.), *Controversial therapies for developmental disabilities.* Mahwah, NJ: Lawrence Erlbaum Associates.

Zechmeister, E. B., & Johnson, J. E. (1992). *Critical thinking: A functional approach.* Pacific Grove, CA: Brooks/Cole.

22

Creative Problem Solving and Troubleshooting

Life is "trying things to see if they work."

Ray Bradbury

When we are practicing our craft as professional behavioral consultants, two repertoires are visible to clients: problem solving and troubleshooting. Our clients would have no need to hire us if they could deal with their problems themselves, and they have no doubt tried all the obvious solutions based on *their* own understanding of human behavior.

What most clients need is a creative solution, and they expect us to be experts in developing these. Identifying the problem, finding a way to measure the behavior, and writing a behavior program are all the initial steps involved in the creative problem solving (Phase 1) that is related to behavioral issues. Once we develop an ethical, workable behavior plan and implement it, we are on to the next phase.

Phase 2 is troubleshooting the plan when it breaks down. Most of the training we provide our behavior analysis graduate students is directed at Phase 1. In classroom and practicum settings, students learn how to pinpoint problem behaviors, measure them precisely, find the cause of behavior problems through observation and functional analysis, and write and implement behavior programs. Although outsiders might consider this a novel approach, for us it is routine. In most cases we can find the function of the behavior and identify reinforcers; the hard part comes in figuring out how to change the contingencies in such a way to modify the behavior without causing harm on a timetable that is reasonable to our consumer and client.

PHASE 1: CREATIVE PROBLEM SOLVING

If your consulting practice specializes in behavior problems, there is a good chance you will see similar behavior problems on a routine basis. You'll be able to develop your own successful patterns of problem solving. For self-injurious behavior (SIB), you might look for escape contingencies or self-stimulation; for classroom disruptive behaviors, you will probably look to accidental teacher reinforcement, peer attention, and escape from academic tasks. Occasionally, there won't be an apparent solution, and your standard modus operandi will fail you. Now what? Fortunately, we can look to other approaches outside of behavior analysis for some help. In *Why Not?* Nalebuff and Ayres (2003) offered a number of strategies from an economist and a lawyer who have been helping business people develop creative solutions for years. Perhaps our field could benefit from thinking differently about some of our behavior problems.

Unlimited Resources

One strategy to employ when you are unable to think of a solution is to step back and ask yourself, "What would I do if I had unlimited resources?" When we can't seem to find the function of a behavior,

> "What would I do if I had unlimited resources?"

we might find it helpful to daydream for a moment about having unlimited resources to do a complete experimental functional assessment. If we had a well-equipped laboratory, plenty of assistants, and a computerized data collection system, surely we could test out enough variables to get to the bottom of the problem. In a setting like this, you could manipulate each variable, scramble the order, replicate conditions, and arrive at a definitive answer. But in the absence of the cold, hard cash, what can you do? One solution is to take your client to a place that does have such facilities. If the behavior is life threatening, as in the case of self-injury, sometimes special funding sources can be found. Or you could bring someone from a lab like this to your setting and request a consult to help you create some approximation to well-controlled conditions. You could also consider Wilder et al.'s (2006) work on brief functional analysis and create controlled conditions of very short durations. The first author once worked with a brain-injured young man, Manny, who had unpredictable aggressive outbursts. Because he was extremely dangerous, he could not live at home, and he was placed in a special locked unit of a mental hospital. He had two-on-one staff for two shifts a day and one-on-one staff at night. This client injured people frequently, including his mother and father, who came to visit him almost daily. Our analysis of the antecedents of these violent behaviors showed that almost anything could set him off. He might misunderstand a comment or a glance or see something he wanted and try without warning to grab it from the person. There was high staff turnover, and Manny was guarded like Hannibal Lecter whenever he was taken off the unit. Manny loved baseball, and we once arranged for him to go to an Atlanta Braves game if he could go a week without aggression. After a few weeks, he met the contingency, and on the big day, he rode to the ballpark without incident. He seemed pleased with himself and was proud that he had overcome great odds. We had great seats and three staff to protect anyone nearby. Manny sat through two innings, grinning and clapping and cheering along with the crowd. He requested a hot dog and a drink, and then, without warning, he jumped up and tried to leave. With spectators watching Manny and ignoring the game, we had to restrain him in the stands and escort him out to the van. We couldn't figure out what went wrong. Why would someone who had the cognitive ability to understand the rules engage in a behavior that would clearly terminate such a powerful reinforcer? In the debriefing afterward, we started talking about the game. Everyone was impressed with how big the players were, how muscled up, how quick, and how powerful. What if Manny was in a different world (instead of being in a locked ward of a hospital) where he couldn't threaten people or push them around, where everyone was bigger and stronger than he is? That day we indulged ourselves for a few minutes and envisioned environments where Manny could cause no harm. Maybe he could be a batboy for a professional baseball team. Or maybe he could be the guy who brings out the new balls for a football or soccer team. What other environments exist where everyone is big and tough and will not tolerate inappropriate behavior? Maybe he could work on a ranch. We actually looked into some of these possibilities. Arranging for Manny to live in such a setting would certainly be expensive, but if you looked at what was being paid for salaries plus room and board at the hospital, plus overhead, this system could actually save money. The closest we came to putting our plan in place was hiring people for Manny's direct-care supervision who were former college football players. In the presence of staff members who were big enough and strong enough to control Manny, he was able to have contact with other people.

Sadly, Manny died before an environment was found where he could be safe from himself. Matching a person to the right environment, regardless of cost, has continued to be a thinking exercise the first author has recommended to others over the years.

Where Else Would It Work?

Another exercise that Nalebuff and Ayres recommended is to take a solution you came up with for one problem and see where else you can apply it. This is the classic "solution in search of a problem."

Tom Coleman and Bill Schlotter, two postal delivery men, were inspired on Halloween night 1987. They saw a trick-or-treater carrying one of those bright, green-glowing cyalume light sticks.

What else could these light sticks be used for? Have you ever considered glowing candy? If you mount a lollipop on top of one of these sticks, the light would shine through the candy, creating a weird and fun effect. Coleman and Schlotter sold their Glow Pop to Cap Candy. Their next innovation was an even bigger hit.

Nalebuff & Ayres, 2003, pp. 31–32

When we were consulting at a developmental center a few years ago, we worked with a client who had profound multiple handicaps (blind, no ambulation, no verbal behavior, could not feed himself) who exhibited severe, periodic SIB. We spent several weeks trying to figure out the causal variables. Suffering one dead end after another, we gave the case to the nurses. "It must be biological. We can't find any environmental events related to his SIB," we said at the end of a protracted case management meeting. The case was given back to the behavior analysis team a week later. At that meeting, a new doctor for the facility was present, and he suggested, "Well, I guess we could do an MRI of his head." It sounded good to us, and the exam was scheduled to occur in a couple of weeks. The good news was they found a likely cause of the head banging: a slow-growing cyst the size of a golf ball in the patient's sinus cavity. The doctors reading the MRI said it appeared the cyst had been developing slowly for quite some time. The remainder of the good news was that it could be removed safely in a routine operation. The immediate effect of the operation, which occurred about a month later, was that the head banging stopped completely. It was so dramatic that we started reviewing all of our other apparently intractable cases to determine if there could be any medical involvement that we (or the nurses) had overlooked. We worked more closely with the medical staff and pushed them harder to make referrals to specialists. An arm scratcher had undiagnosed allergies, a rectal digger had hemorrhoids, and another head banger actually had something stuck deep in his ear canal. These cases teach an important lesson: Always, always consider the possible biological factors in behavior.

Flipping It

One final strategy that Nalebuff and Ayres (2003) suggested that seems like it might be useful to behavior analysts who are looking for a creative solution is an exercise called "flipping it." In this exercise you basically consider what would happen if a standard product or service were reversed or flipped over. Consumers saw this in recent years when creative designers at Heinz finally figured out a way to help us get the ketchup out of the bottle. They redesigned the container so that it sits on the table upside down! In behavior analysis, we consider the contingency involving contingent reinforcement to be central to the way that behavior change is engineered. What if we flipped this around so that reinforcers were noncontingent? Actually, this has been demonstrated to some extent in the research literature, but it does not seem to have made any headway in clinical and educational applications. If an individual will work really hard and undergo a lot of pain to get a reinforcer, why don't we just give it to him and see what happens? We actually have a concept to explain how this works: the Establishing Operation (EO) or, more recently, the Motivating Operation (MO). By reducing the power of the EO/MO, we greatly reduce the strength of the behavior, perhaps to the point where we could again begin to use the reinforcer contingently to shape an appropriate behavior.

What about other standard operating procedures for behavior change, such as when in a classroom we try to train a teacher to give reinforcers contingent on a disruptive student's quiet behavior? What if we prompted the child to reinforce the teacher? Or what about an unruly child who is sent to the office? Instead, what if the child reported to the office first thing in the morning and got his assignment there, and if he completed it, he could visit his classroom for a short time? On one consulting assignment, there was a young man, Marc, who wouldn't pull his pants up for anyone or anything. As a part of his program, he lost privileges if he went around without his pants or with his pants down. As a result, he never got to go on any outings. One day, a new bus driver who didn't know about this rule took Marc to the mall with six other moderately mentally retarded clients. When the driver and clients got back, the staff rushed the driver and shouted, "What happened?" The driver said, "Nothing. Why?" When told about Marc's program, the bus driver looked stunned and said, "Marc wasn't one bit of a problem. When we got to the mall, he hiked up his pants, and off we went. I didn't even prompt him."

PHASE 2: TROUBLESHOOTING

The fun really begins when your program plan is finally approved. Now it's time to start training all the people involved to play their new role as a *junior assistant behavior-change technician.* Because we work with the model that behavioral interventions are best done in the natural environment by those who live there and work with the clients, this is where we need to look when the program flops. These indigenous individuals know your client well and can often yield the information needed to redesign the key elements. Our helpers are indispensable, invaluable, and essential to the behavior-change effort. They can also be unreliable, inconsistent, and sensitive to minor setbacks that are as inevitable as a drought in the Sahara. We greatly underestimate how much training and support our junior assistants need, and the first form of troubleshooting involves looking at these helpers. In one school, we, on a referral from the principal, worked with Ms. Carlton, a first-year third-grade teacher who had no control of her classroom. Observations showed us that she was unaware of what was going on around her, that she attended to inappropriate behavior consistently, and that she had virtually no positive affect in her voice. After some consultation, she decided she was willing to try a reinforcement system, even though she was reluctant. Ms. Carlton was given instruction in the use of reinforcers. She seemed to grasp the concept, had no questions, and appeared ready to go. Everything was going well. The principal funded the reinforcers, and Ms. Carlton was to begin the new program on the following Monday. We checked in on her the following Friday just to make sure everything was going well. There was noise coming from down the hall, and things in Ms. Carlton's classroom had gotten worse. The behavior analyst sat in the corner of the classroom and waited for class to end. Ms. Carlton started in a flat steady voice,

> I decided to use candy instead of tokens. It made more sense to me than having them earn tokens and then my having to hand out the candy too. Plus, my husband said the tokens were too confusing. But after the first day, I hadn't given out any candy, and the kids started complaining because I told them in the morning that I would. So the next day, I just decided to put the candy dish right here on my desk. I told them, "When you have completed your assignments, you may come up quietly and take one piece of candy." Well all the candy was gone by 11:00, and the next day it was gone by 10:00, so I just said this reinforcement stuff isn't working and stopped it. I tried, but it really is too much trouble when you are busy teaching.

What Can Go Wrong?

As described earlier, most behavior plans have a lot of moving parts, and a lot of things can go wrong. People can forget to give prompts, they can be inconsistent in the delivery of reinforcers, or they can pair reinforcers with an unpleasant look or a sarcastic comment. Sometimes your chief assistant, a teacher aide, for example, just doesn't show up, and you have no backup.

"As described earlier, most behavior plans have a lot of moving parts, and a lot of things can go wrong."

In one classroom, we set up a great token economy as a pilot and were prepared to expand it to the entire sixth grade when we discovered that an enterprising student made counterfeit tokens and was selling them on a black market in the cafeteria. In another case, a mom who was trained quite carefully to use backward chaining with her child, starting with shoe tying, gave up after a couple of days and decided her mildly handicapped daughter could wear flip-flops to school. A gym teacher who was shown how to use a modified time-out procedure to manage disruptive behavior in his class was seen standing over a bench full of middle school youths, berating them for not following instructions to "sit quietly and not talk," while the rest of the class was out of control on the soccer field. Basically Murphy's Law applies to *all* behavioral interventions: "If anything can go wrong, it will." Behavior analysis is second nature to us but is a foreign language to almost everyone else.

Troubleshooting Tip 1

Train your helpers to criterion. Do not assume that just because you've demonstrated how you want something done and asked, "Any questions?" that you are done. Even if no hands go up when you ask, "Any questions?" you can bet that someone is confused. Training should follow this order: Describe what you want and the rationale for it, demonstrate, have the helpers practice, give feedback, practice some more, give more feedback, and repeat until the helpers are fluent with their new routine.

Troubleshooting Tip 2

The first day that your treatment is to start, *be there.* If you let your new intervention start without your being present, you will have to rely on hearsay, which is guaranteed to be less than adequate.

Troubleshooting Tip 3

Always debrief after the first day and the second day and the third day. Watch your helpers closely as they describe how they think it went. You will be reading body language to see if they felt comfortable in their new role or perhaps felt ill at ease with the tasks that were assigned to them.

Troubleshooting Tip 4

Your behavior plan should also be evaluated by the data that are coming in each day. Graph the data every day, and be critical of your own work.

Troubleshooting Tip 5

Always have a Plan B.

SUMMARY

Creative problem solving and troubleshooting are essential skills for the successful behavior analyst. These are skills that require attention to detail, persistence in looking for alternative ways of changing behavior, and diligence in following up with any behavior plan. Most plans will not be as perfect as they are initially written to be. Your ability to remain calm, reconsider your basic assumptions about the case, and come back with an improved plan are the bedrock proficiencies you will use every single week that you practice as a behavioral consultant.

FOR FURTHER READING

de Bono, E. (2008). *Creativity workout: 62 exercises to unlock your most creative ideas.* Berkeley, CA: Ulysses Press.

Michalko, M. (2001). *Cracking creativity: The secrets of creative genius.* Berkeley, CA: Ten Speed Press.

Nalebuff, B., & Ayres, I. (2003). *Why not? How to use everyday ingenuity to solve problems big and small.* Boston: Harvard Business School Press.

Silber, L. (1999). *Career management for the creative person.* New York: Three Rivers Press.

23
Understanding and Using Power

We have learned that power is a positive force if it is used for positive purposes.

Elizabeth Dole

*P*ower certainly has a negative connotation in our culture. It suggests manipulative people doing nasty things to defenseless innocents in unconscionable ways. Certainly our history, especially our recent history, brings this to mind in dramatic fashion. The fictional Gordon Gekko (from Oliver Stone's 1987 movie *Wall Street*), deceased Saddam Hussein, and still influential though reclusive dictator Kim Jong-il all are at the top of our list of corrupt, disdainful, powerful people.

In business, however, the term *power* is defined more objectively as "the potential to allocate resources and to make and enforce decisions" (Harvard Business School Press, 2005, p. xi). For behavior analysts working in organizations, agencies, or consulting firms, learning about power is essential if they want to be effective and have influence in their chosen field. You may not aspire to a powerful position, but if you do not at least understand power in the workplace, you can be at a significant disadvantage. Ordinarily, the progression through one's career does involve moving up, taking on more responsibility, and having a greater influence on how resources are distributed. Salary increases, greater recognition, and certain professional trappings accompany these job-related advances, and thus there are built-in incentives for behavior analysts to acquire the skills and attitudes associated with a greater span of control and influence.

"Ordinarily, the progression through one's career does involve moving up, taking on more responsibility, and having a greater influence on how resources are distributed."

SMALL-SCALE POWER

Much of what is written in the business literature about power has to do with how very large corporations operate. Multiple departments vying for resources or competing with one another for the attention of the board of directors or CEO are common power struggles at the corporate level. With big money at stake, the contingencies can bring out the worst in people.

At their current stage of development, behavioral consulting firms or agencies providing behavioral services in schools or residential settings rarely have more than 100 employees. Most are small businesses or nonprofit organizations with 50 or fewer people. Nonetheless, there is a power structure in these smaller organizations. If you are the new hire in a small consulting firm, you are well advised to remain alert and pay attention to the informal rules of the game, how decisions are made, and who calls the shots. Someone in the organization has to determine how much people are paid. Chances are this person or department also develops and maintains the resources necessary to run the company. Someone else has to resolve disagreements between employees, serve as the face of the organization to the community, or motivate employees to support a new company goal. Each of these roles creates an opportunity for an individual to gain power.

SOURCES OF POWER

Job Title

When you first join an organization, you will usually start at the bottom and will have a title such as therapist I or consultant, and you will have a supervisor with a title such as manager, director, or administrator. The people in supervisory positions confer a certain amount of power in that they can set your work schedule, assign you clients, or prescribe what you can and cannot do. Not all people in this position are effective, and many give away any semblance of authority just by the way they manage people. This is where being a good observer and understanding how to apply your behavior analytic skills (Chapter 13, "Think Function") comes into play. Managers who are petty and arbitrary in their decision making erode their own power to engage and influence people at work. On the other hand, those who take this position and use it as an opportunity to shape behavior and "bring out the best in people" (Daniels, 2000) can make a big impression on company officials. The title alone does not confer much power; it only creates an opportunity for the behavior analyst to use her shaping skills (Chapter 14, "Use Shaping Effectively") to make a positive impression on management. When you have advanced into a basic supervisory position, your job title begins to confer more authority. Being a director, department head, or senior consultant gives the behavior analyst an opportunity to get involved with strategic decisions that influence policy, to detect opportunities to grow the company, and to play a significant role in employee selection and training. This step up the corporate or organizational ladder bestows considerable power on an individual to possibly hire and fire people and to direct company assets to certain preferred projects. It might take 5 years to get to this level of leadership in an organization, and those who wield power as a blunt instrument are usually weeded out. Upper-level management does not take kindly to managers who use their power to baldly advance themselves or abuse those under them. Behavior analysts in particular should do well in positions of authority because they understand that the higher they move on the organizational chart, the more powerful reinforcers they have to dispense (Chapter 16, "Performance Management").

Relational Power

As described in *Power, Influence, and Persuasion* (Harvard Business School Press, 2005), a second form of effective influence comes directly from your affiliation with other individuals in your organization. By forming coalitions with like-minded others, you can have more influence than you would have individually. This begins with effective networking to learn who your colleagues are and how they feel about certain issues. By joining these colleagues on projects of common interest, you can begin to bond with them, and when you need their support, you can ask for it and count on them to deliver. Behavior analysts who are sensitive to the needs of others and provide the backing co-workers need to accomplish their objectives are usually welcomed as colleagues. Reciprocation is the unspoken currency that drives effectiveness in most organizations. Watch for these opportunities, and you should find yourself moving quickly along on the management track.

> "Reciprocation is the unspoken currency that drives effectiveness in most organizations."

Personal Power

One final source of influence in most organizations is related purely to the personal characteristics that make you a desirable partner under all circumstances: trustworthiness, expertise, charisma, accomplishments, enthusiasm, and self-confidence (Harvard Business School Press, 2005). Lacking these traits, you may find it difficult to move into a position of leadership and influence, but knowing early on that they are important gives you time to work on self-improvement tasks if necessary. This self-improvement can take some serious self-examination. Many people will shrink from this task, believing that their personality is set, and they should be accepted just as they are. Behavior analysts should not allow anything to be a barrier to self-improvement. We know how to pinpoint key skills and how to monitor and track success. We can do this with clients, and we can do this to improve our own performance.

Soft Power

The blatant use of power to force people to comply with requests is incompatible with how behavior analysts should relate to people. If we are fully activating our technology of behavior change in our

> "If you have established yourself as a powerful reinforcer with your subordinates and if you make clear what you want them to do, make sure they are capable of doing it, and provide intermittent reinforcement, you'll never have to raise your voice or threaten or cajole them."

everyday lives, we should be subtle in our attempts to influence others so they want to respond to reasonable requests. If you are in a position of power or you are a reinforcing person who is well liked, others might be eager to comply with your requests in order to impress you or make you happy. All of this seems to describe what is meant by "soft power" (Bixler & Dugan, 2001). If you have established yourself as a powerful reinforcer with your subordinates and if you make clear what you want them to do, make sure they are capable of doing it, and provide intermittent reinforcement, you'll never have to raise your voice or threaten or cajole them. You won't need to dominate a meeting or demand attention, loyalty, or respect with harsh words or intimidating comments. A really powerful behavior analyst should be able to quietly describe her goals and objectives to a room full of associates; ask them for input, suggestions, and ways to accomplish the goals; and, with subtle head nods, smiles, and soft descriptive compliments, get an unspoken pledge of support and commitment. As a behavior analyst, you will no doubt have experienced quite the opposite at some point: a loud, abusive bully who demands respect and conformance and who allows no ideas but her own. You can use this type of person as a benchmark for the other end of the continuum.

Who Are You? What Are Your Goals?

The topic of power and how it is used in organizations makes many behavior analysts uncomfortable. It is as though the concept of power should not be discussed in polite company. We believe that behavior analysts can and should play a larger role in guiding organizations and influencing society. We think that we have an extremely important message to deliver: Behavior analysis can be a powerful force for good in our culture. Only by understanding how human behavior works can we begin to produce cultural contingencies that support cooperative, nonaversive, productive behavior (Skinner, 1957). We have to get the word out, and one way to do so, clearly, is for behavior analysts to become part of the power structure in organizations, corporations, and society. There is no reason that behavior analysts should not be on the school board of their community or participate in other activities of state and local government.

In some settings, we look around and see people from other professions having influence, taking positions on important topics, and acting as though our role is limited to behavioral technicians. Too often, others get to set the agenda and call the shots; we get to listen and follow their lead. If we establish early in graduate school, however, that we expect our students to become the leaders of the future and encourage them to study the practices of powerful people, we should be able to turn around this sad situation.

As you begin to understand how power is acquired and what the benefits are of having it at your disposal, you should, we hope, engage yourself in a dialogue of discovery about issues larger than simply working with clients or consulting with parents, teachers, or CEOs. The technology about which you are acquiring knowledge gives you a great deal of power. You have expertise that very few people have, which is an understanding of why people do what they do. Countless books providing theological, cognitive, sociological, and other explanations for behavior fill the bookshelves of bookstores. Because they cannot be observed, measured, or replicated, none of these explanations come close to the evidence-based science of behavior that we have achieved over the past 40 years.

Why have we not had greater influence? We believe it is partly because individually we have neglected the study of power and have not encouraged our students to take the steps necessary to acquire it

> "In its simplest form, power is nothing but the use of reinforcers to change behavior."

and make use of it. In its simplest form, power is nothing but the use of reinforcers to change behavior. A *powerful* contingency is one that is highly likely to produce a given behavior. We know how to design these types of contingencies. What can you do to help develop this part of the science? Are you prepared to do more at work than to simply do your job? Can you see yourself as a decision maker, someone who has the vision to see a different future for your company or your community? This is going to happen only if you can harness the energy and effectiveness of colleagues around you. Can you begin to manage using your behavior to influence others so that everyone in your department is as engaged, enthusiastic, and committed as you are? This is the first step toward developing a personality that your friends and colleagues will want to emulate.

SUMMARY

Power is the "the potential to allocate resources and to make and enforce decisions" (Harvard Business School Press, 2005). A close examination of the ways that power can be acquired shows that the process of acquiring power is entirely behavioral and that there is little mystery involved. Learning to acquire power by moving up in your organization, developing coalitions of allies, and nurturing your own desirable leadership characteristics are essential if you are to become an effective behavior analyst consultant. Having learned how to use power, you should be able to add considerable value not only to your organization but to your community as well.

FOR FURTHER READING

Bixler, S., & Dugan, L. S. (2001). *5 steps to professional presence.* Avon, MA: Adams Media.

Harvard Business School Press. (2005). *Power, influence, and persuasion.* Boston: Author.

Skinner, B. F. (1957). *Verbal behavior.* Englewood Cliffs, NJ: Prentice Hall.

24

Training, Coaching, and Mentoring

You get the best effort from others not by lighting a fire beneath them, but by building a fire within.

Bob Nelson

Training, coaching, and mentoring are three essential strategies for changing human behavior in the workplace. In some cases, you will be working with mentors, giving them the skills they need to be a part of your behavior-change team. On other occasions, the goal of coaching will be to improve an individual's attitude, technical expertise, general effectiveness, or social interaction skills. And at some point in your career, a junior colleague might ask you to serve as a mentor. Each role or method is appropriate under certain circumstances, and each one embodies all of the elements of behavior analysis. Whether you are a graduate student just learning the ropes or a senior behavior analyst, learning more about how these behavior-change strategies are used will make you a more effective professional.

TRAINING

"The success of applied behavior analysis (ABA) is completely dependent on modifying the behavior of mediators, such as staff, peers, and parents" (Sturmey, 2008, p. 159). Nothing could be truer, and our success depends on a thorough, fluent knowledge of available training techniques. We generally think of training as consisting of a packaged set of skills taught in a standard way to frontline mediators. These mediators may consist of direct care staff, parents, foster parents, teachers, teacher aides, and others who deal with behavior management and instruction all day long. The training of these mediators has been routinized, and because of either great demand or high turnover, specialized trainers often do this work.

Before you get to training, you have to be sure that this is the right solution to a performance problem. All too often it is assumed that any deficit behavior has training as the answer when in fact performance management (Chapter 16) is more appropriate. Having determined that training is the answer, it is probably a good idea to ask, "How would I know if the training worked?" Most prepackaged training workshops are put together, implemented, and then put on the shelf until the next group comes along. Our approach is determining in advance if training is needed (we always take a baseline). Kirkpatrick (1994), who advocated evaluating the training once it has been delivered, also used our approach in instructional technology.

This, of course, raises the question of how training is evaluated. Kirkpatrick's model consists of four levels. Level 1: How do the trainees like the training? Level 2: Have the learners acquired any skills? Level 3: Have the trainees generalized their new skills to the environment where they are used every day? Level 4: Has there been any demonstration that the new skills now shown in the home, classroom, or business environment result in any effect on behavior change or the bottom line? Levels 3 and 4 represent very big challenges to behavior analysts who work as trainers or supervise training enterprises. When it comes time to review best-practice recommendations, these standards should be adopted for our field.

Traditional Training

Organizations were training staff long before applied behavior analysts came along. In general, traditional staff training models do not share our respect for taking a baseline before training begins to determine how much behavior acquisition has occurred, and they don't follow up to determine if training is effective. "It was required, it was done, it is over" is the standard response one gets if one inquires about training policy and procedure. Traditional training almost always involves group instruction in a classroom-like setting. It consists of stand-up training, some handout materials to supplement the workshop, and a sign-out procedure that holds everyone who attended accountable for what they heard (whether or not they actually understood it). The assumption of this style of training is that trainees can listen to a presentation and perhaps even view a video clip and then translate it into an appropriate response to be delivered at just the right time with a client. I know, you're saying, "Surely not?" but yes, it's true. This is the assumption, and this is the method that has been used in education, mental health, developmental disabilities, rehabilitation, and vocational settings for at least 30 years.

Behavioral Training Model

Our model provides quite a contrast. Behavior analyst professionals involved in training have a different tradition: We emphasize the need for evidence for what we do, and this includes our training practices. Often referred to as "behavior skills training," our approach typically involves four distinct components.

Step 1: Instructions and Motivation   Behavior analysts keep verbal instructions to a minimum, which is in contrast to traditional training, and in this phase, they will describe the rationale for the training, outline the steps of the skills to be acquired (showing this visually in a task analysis), and answer questions prior to moving to Step 2. A primary objective of this phase is to motivate the learner to want to participate. Describing success stories from previous trainees or clients who have benefited from training is one of the methods used for motivating learners. The trainer needs to be a peppy person with a lot of excitement in his or her voice and mannerisms that relate well to the trainee group.

Step 2: Modeling   Although a lot of what we teach our mediators involves verbal behavior (ideas about the immediacy of reinforcement, how extinction works), the real delivery system requires a person who can *perform* correctly (motor behavior) under just the right circumstances. The acquisition of motor skills starts with seeing how the behavior is to be carried out, that is, modeling the performance. Ideally this would be done under the actual circumstances in which the behavior is required. This might be difficult or impossible, so a fall-back position might be either a role-play with a confederate or a videotape (or CD or clips available on the Internet). It is important that the behavior being trained is clearly defined and presented. It should also be separated from other skills so that trainees don't become confused. Each step that is being modeled should match the task analysis from Step 1.

Step 3: Practice With Feedback   One distinguishing feature of the behavior skills training model is that trainees are required to *demonstrate* the skill modeled in Step 2. It is here that the trainer learns if his training method is actually working. Approving head nods and smiles are no match to actually seeing the students demonstrate the new behavior. And they can't get by with just doing it once. In effective training, trainers should demonstrate the required behavior repeatedly until it is clear the students are fluent. Because what we are teaching people is a skill to be emitted under certain specific circumstances, the trainer needs to include both S^Ds (discriminative stimuli) and S^Δs (pronounced "S-Delta" and sometimes referred to as an extinction stimulus, this is a stimulus that sets the occasion for a decrease in operant responding); that is, trainers need to know exactly when to apply a positive reinforcer and when *not* to. Practicing until trainees are fluent usually takes longer than traditional training, so the trainer has to use all of her social and reinforcement abilities to keep everyone motivated. Many people are hesitant to engage in training like this because of the potential exposure to ridicule from others in the class if they do something wrong. Another requirement of a good trainer is the ability to manage the class in such a way that this does not happen.

Step 4: Follow-Up, Corrective Feedback, Maintenance   As quickly as possible, the trainer needs to arrange for the students to return to their settings and engage in the newly acquired behaviors. The trainer, not the students' supervisor (although it is desirable if this person were one in the same), needs to set aside time to observe the individuals use the new skill and provide the necessary reinforcement and corrections (if necessary). Ideally, the trainer would have observed the trainees before the training class and would be able to see a significant difference in performance. This difference in performance is the most significant form of reinforcement for a trainer and can sustain trainees for a very long time. Having seen that the newly trained people can perform admirably, the trainer should next meet with the supervisor, report on the success, and describe how the supervisor can take on the next phase of training. This phase involves maintenance of the skill. Behaviors that do not have automatic feedback built in require systematic observation by a dedicated supervisor along with intermittent reinforcement (see Chapter 16, "Performance Management").

> "Behaviors that do not have automatic feedback built in require systematic observation by a dedicated supervisor along with intermittent reinforcement."

COACHING

Ray was a Board Certified Behavior Analyst® (BCBA) with 2 years of experience in developmental disabilities and "good" ratings from his supervisors as a consultant. Denise, his supervisor, was not completely happy with Ray because she had to solve problems caused by his somewhat rough style of interaction. Ray came across as rather blunt, although when given feedback, he said, "I was just being honest with people. I like to keep it real." Ray would be turning 30 in about 6 months, and it seemed to Denise that it was time for him to "grow up and act like an adult" when he was around his clients, colleagues, and other professionals.

This scenario is ideal to set the stage for a coaching intervention. The behavior is narrowly defined, has an impact on the clients, and is worrisome and bothersome to the supervisor. Denise could just give Ray a low rating on his next performance review, which was coming up in a couple of months, but it occurred to her that coaching might be the answer. Because she was a BCBA-D (a doctoral-level BCBA) with 10 years of supervisory experience, Denise had the right credentials, and she was certain she could do this.

In business settings across the country, coaching is used with increasing frequency. This is possibly a result of there being an increased recognition that traditional performance appraisals are too much of a blunt instrument. Furthermore, the classic annual review is too delayed to have much impact. Counseling has been an option for years, but it doesn't have much credibility for a situation like this. Coaching is generally described as having four steps (Harvard Business School Press, 2004).

Step 1: Observation

Because coaching is a behavioral process, it obviously must start with direct observation of the individual. You cannot work from hearsay or off-the-cuff remarks overheard in the break room. Denise needs to arrange to observe Ray's interactions with clients and colleagues in the settings where they naturally occur. She will need to take notes not only on performance gaps and skill deficiencies but also on highly valued skills that he does have in his repertoire.

Step 2: Discussion

Prior to the actual coaching, Denise needs to sit down with Ray and have a discussion about her concerns. She should ask Ray to give his own self-assessment. If Denise detects an intransigent attitude on Ray's part, she may decide to fire him rather than take the time to try to rehabilitate his repertoire. The goal of this step is to bring to Ray's attention what the supervisor sees as the problem areas. She will listen to his version and then attempt to persuade him to participate willingly (remember Chapter 9 on persuasion) in the process. Positioning Ray so he sees this as a useful exercise that will enhance his performance as a consultant and possibly lead to advancement in the organization is a fundamental goal of Step 2. Coaching should be seen not as punitive or threatening but rather as an objective, behavioral method of increasing the employee's effectiveness.

Step 3: Active Coaching

This step begins with an agreement between the two parties that coaching is needed and that certain goals and objectives are understood. This should not be a problem for Denise because she is a good observer, has good detailed notes, and can articulate a set of observable behavior-change criteria for Ray. With Ray's concurrence, the actual coaching can begin. Ordinarily, the coach would start with a behavior that is obvious and, in her estimate, fairly easy to change. Denise wants to start with some changes that Ray can make without a lot of trouble, which gives her an opportunity to be very reinforcing and show that she is not trying to be punitive. Denise might model and role-play certain scenarios with Ray. The role-play training will be timed to occur just prior to an opportunity Ray will have to practice the new skill. The coaching will occur in small amounts over several weeks and will involve discussion, role-play, practice, feedback, and a chance to apply the skill. Each cycle will take on a slightly more difficult problem, but as Ray experiences some success, he should gain the confidence to continue working with his supervisor. Denise will also benefit from this experience because she will be able to see immediately whether her coaching is working, which is her main reinforcer for this whole exercise.

Step 4: Follow-up

Behavior analysts will find that coaching is a natural way to solve problems with individual employees. Coaching can also be used to upgrade the skills of those on the team who could make a greater contribution if they could take on some additional responsibilities. The follow-up phase after the coaching is a natural part of what behavior analysts do. As a behavior analyst who is providing training to others, you'll want to know if the natural contingencies in your work environment can maintain skills. You should be prepared to modify the environment if you need to add reinforcers that will maintain behaviors. In the business world, the essential feature of maintenance of behavior is not well recognized, and any behavior analyst involved in coaching should add this maintenance to the coaching checklist.

MENTORING

As a new member of the applied behavior analysis club, you may find that you sometimes feel lost or alone. You might have questions that your other junior colleagues can't seem to answer. You might feel that you are destined to have a

> "As a new member of the applied behavior analysis club, you may find that you sometimes feel lost or alone."

greater impact on your company or organization, but you're just not sure what moves to make or how to make them. You may not know this, but what you are looking for is a *mentor*, a person with useful experience, skills, and expertise who can offer advice, information, or guidance to help advance your personal or professional development (Harvard Business School Press, 2004). The *protégé* is the person who initiates the search for a mentor who will establish a long-term relationship with the protégé and provide a variety of supports ranging from opening doors to providing career counseling. In some cases, the mentor may provide protection for the protégé in difficult political interoffice situations or serve as a sounding board for new ideas. Mentors challenge protégés to expand their horizons and develop their capabilities and potential as consultants. The mentor is very familiar with the skill set of the protégé and may even recommend that the protégé consider other lines of work where the PhD or MBA is required.

Entering into a mentoring relationship is no small decision. The time required can be enormous, so the mentor must consider whether the investment is worth it. A potential protégé has to be qualified, as in whether she is hungry for knowledge, ambitious, and eager to learn. A mentoring relationship can continue for 2 to 5 years initially, so both parties must have matching, compatible personalities. Needless to say, the mentor has to be a great behavior analyst who feels ready to give back to a younger person, to bring the young behavior analyst along, and to help advance a career.

SUMMARY

Training, coaching, and mentoring are three strategies for behavior change. These strategies are used in a wide range of circumstances from group instruction to career counseling. To be effective as professional consultants, behavior analysts need to be proficient in training and prepared to serve as a coach when the time is right. Younger behavior analysts may want to consider seeking out a mentor if they feel the need for career advice and counseling.

FOR FURTHER READING

Harvard Business School Press. (2004). *Coaching and mentoring.* Boston: Author.

Kirkpatrick, D. L. (1994). *Evaluating training programs: The four levels.* San Francisco: Berrett-Koehler.

O'Neill, M. B. (2000). *Executive coaching with backbone and heart: A systems approach to engaging leaders with their challenges.* San Francisco: Jossey-Bass.

Reid, D. H., & Parsons, M. B. (2002). *Working with staff to overcome challenging behavior among people who have severe disabilities.* Morganton, NC: Habilitative Management Consultants.

Sturmey, P. (2008). Best practice methods in staff training. In J. K. Luiselli, D. C. Russo, W. P. Christian, & S. M. Wilczynski (Eds.), *Effective practices for children with autism.* New York: Oxford University Press.

25

Aggressive Curiosity

The important thing is not to stop questioning.

Albert Einstein

W e begin this chapter on aggressive curiosity with comments by Jon Bailey.

On Sapelo Island in southwest Georgia, only accessible by ferry or private boat, is the community of Hog Hammock. I want to go there someday. I'm curious about this place because there is a historic 200-year-old West African slave burial site called Behavior Cemetery. If I can make my way there, I understand I can track down a fisherman named Maurice Bailey, and he might give me a guided tour. With luck, I might be able to talk to Cornelia Bailey (2000), who wrote a history of Hog Hammock. I'd like to learn more about how this Gullah-Geechee community of slaves came to give this unusual name to the cemetery. Who knows, we might even be related.

Moving on to another part of the country and another topic, I would also like to visit with Joshua Klein someday. He has an amazing presentation on TED.com in which he demonstrates a vending machine for crows.* He trained crows to pick up lost change in exchange for peanuts. Klein doesn't call what he does operant conditioning, and he doesn't seem to know there is a whole international network of behavior analysts who share his passion for understanding behavior.

---

* See ted.com/index.php/talks/joshua_klein_on_the_intelligence_of_crows.html.

I am fascinated with behavior and have been since my introduction to the topic in 1961 in Jack Michael's Introduction to Behavior class. Jack had a way of telling stories. When he talked about discoveries in the animal lab and his interactions with Ted Ayllon, he captured my imagination. These were fascinating stories of scientists making discoveries about how behavior works. I was amazed to learn that behavior could be observed and measured systematically and that it could be changed. People's lives could be improved just by changing the contingencies of reinforcement. This was a radical departure from the traditional thinking that people did what they did because they wanted to or had some genetic trait that made them aggressive, submissive, or manipulative. Jack instilled in me a curiosity just by the way he enthusiastically described the settings, the people in the settings, and how he and Ted thought about how to help them. Jack and Ted weren't working with much. They had B. F. Skinner's *Walden Two, Science and Human Behavior,* and *Verbal Behavior* and a few volumes of the *Journal of the Experimental Analysis of Behavior.* But they were unencumbered by tradition, they were breaking new ground, and they knew it. They moved the field forward with their aggressive curiosity. It was an exciting time.

To be an effective behavior analyst requires an abiding curiosity about people in exotic and ordinary settings. What causes a young woman to strap on a bomb and head into a crowded market in Iraq, killing 38 innocent civilians?

> "To be an effective behavior analyst requires an abiding curiosity about people in exotic and ordinary settings."

How did a U.S. Airways pilot stay absolutely calm and safely land a crippled plane in the Hudson River, saving 155 lives? The histories of reinforcement leading up to these respectively depressing and uplifting performances deserve study. As behavior analysts, we should be curious about these kinds of behaviors. We could probably learn a great deal about basic behavioral processes if we knew more about how terrorists are selected and trained, and we might be able to save lives with this knowledge.

There are interesting phenomena all around us worth thinking about, problems to solve, and incredibly interesting ideas to consider from a behavioral perspective.

To stimulate your curiosity, you need to start with a good source. One of the very best sources is the *New York Times Magazine*'s "Ideas and Trends" column that appears regularly. In mid-December, the magazine (a supplement to the Sunday *New York Times* newspaper) publishes the *Annual Year in Ideas* in which inventive, creative schemes are described.* These can solve a whole host of unusual problems, both real and fictional. Here are just a few, with some behavioral commentary.

Air bags for the elderly: Many older people have a fear of falling that is not unjustified. It has been reported that falls are the leading cause of death in those who are older than age 65 (BBC News, 2008). A company in Japan came up with an inflatable wearable air bag that looks like a fishing vest and can inflate in a 10th of a second when activated by a motion sensor. The price of $1,400 is out of reach for most people. Could there be a cheaper behavioral solution? What if older people who were inclined to fall wore just the motion sensor, and it gave them feedback on their posture? O'Brien and Azrin (1970) dreamed up this solution, under the general category of behavioral engineering, back in 1970. Maybe it's time to revisit this approach for other behavioral issues.

Goalkeeper science: Israeli scientists have analyzed the behavior of soccer goalies and discovered that "94 percent of the time the goalies dived to the right or left," even though their best move would have been to stay in the center, that is, they should have done nothing. The authors generalized their findings to the strategies of corporate CEOs during turbulent economic times. CEOs tend to suddenly change course when the wisest course of action might be to stand firm. If behavior analysts were working in Fortune 500 companies, they might be able to detect these changes in behavior

---

* *New York Times Magazine*, December 14, 2008.

and devise monitoring and feedback systems based on previous historical trends.

The one-room school bus: Many children spend up to 3 hours per day riding a bus to and from school. This is clearly wasted time. Furthermore, some of our behavior analytic research has indicated that long bus rides produce some fairly serious behavior problems for the drivers to deal with. Dr. Billy Hudson, a professor at Vanderbilt University, came up with an idea to transform the bus, now wired for the Internet, into a mobile classroom. Students from Grapevine, Arkansas, can enroll in online courses and do the work on their laptop computers. If this initiative were tied in with behavior analysts who work in the schools, they could additionally evaluate the effects of this program on behavior both on the bus and later at school. A curious behavior analyst might ask what other applications there could be for the concept of mobile learning opportunities. A behavior analyst with *aggressive curiosity* would contact Dr. Hudson at Vanderbilt and arrange to visit the program. The behavior analyst would then return home and work feverishly to determine if the technology could be replicated with the local rural school system where the most frequent referrals for behavior programs are children who have long daily bus rides.

CURIOSITY VERSUS AGGRESSIVE CURIOSITY

Although regular garden-variety curiosity is fairly common in our culture, newspaper readership is on the decline. This is because people are getting their news faster and from a wider variety of sources than ever before ("Newspaper Circulation Rising Globally, Down in U.S.," 2008). Web sites stimulate curiosity by having hot links embedded in the article you are reading so you can find related stories and background data. The response cost for quenching your curious thirst for ideas and information has gone way down. Theoretically, this should reinforce your search for new ideas. If you have *aggressive curiosity*, you have such a driving need to understand how things work that you are willing to devote some considerable time on your search for an answer, and then you actually take some action. As a behavior analyst, you will serve your client far more fully if you become completely immersed in his situation.

One new behavior analyst was given an opportunity to work with a safety improvement systems consulting firm in Australia. This assignment could be daunting if you knew you would be working in steel, coal, gas, and aluminum manufacturing plants and had never even seen one up close. After 3 months of digging day and night for information and reading safety manuals and company documents, our behavior analyst reported for duty so well prepared that she knew more about most of these heavy-industry plants than many of the management staff. Her newly acquired expertise, fueled by aggressive curiosity, carried her right into the heart of the consulting opportunity and the adventure of a lifetime.

HOW TO INCREASE YOUR AGGRESSIVE CURIOSITY

If you are moderately curious about your culture and the human behavior driving it, and resulting from it, and want to go to the next level, here are some activities for you to consider.

Read Broadly

One way to increase your curiosity is to read what other people in other professions in other parts of the world are doing. You need a lot of stimulation, many things to think about and ponder, and issues to relate to one another to activate your curiosity. How did they come up with this idea? Why do they deal with this problem in this way? You will find yourself saying, "If they hired me, here's what I would do to solve that problem." If you do this routinely and get into the habit of talking back to your newspaper or computer screen, you'll find that ideas jump out that relate to a problem you're having right now in your organization or with a client.

Keep a Notebook

You will come up with ideas throughout the week based on something you heard on National Public Radio (NPR), seen on the Internet, read in the paper, or picked up in a conversation. If it is new or strikes you as unusual or funny, jot it down and indicate the date and maybe where you heard it so you can track it down later. The first author keeps a small notepad and pen in his car and will pull off the road to make a note about something he heard on *All Things Considered* on NPR. This was the beginning of a string of behaviors that started as an interview with Amy Sutherland, the author of *Kicked, Bitten, and Scratched* (2006). The phone interview led to her accepting an invitation to give the keynote address in 2008 at the annual meeting of the Florida Association for Behavior Analysis. She stimulated the thinking of nearly a thousand behavior analysts with her talk that was based on her 2008 best-selling book *What Shamu Taught Me About Life, Love, and Marriage.*

Watch Indie Films

The blockbuster films that occupy 90% of the screens in America don't often provide us with food for thought; independent films do. Often made on low budgets by creative people with a very unique

> "The blockbuster films that occupy 90% of the screens in America don't often provide us with food for thought; independent films do."

perspective on the world, these films portray human behavior in all its vast and glorious diversity. They capture human behavior under circumstances we could never imagine and challenge us to understand how and why people do what they do.

Meet New People

If you have a regular posse and do everything together, you will soon begin to think in such uniform terms that you'll begin finishing each other's sentences. Every now and then, consider making a new friend from a different line of work or a different religion or culture. Use this as an opportunity to see how this person reacts to challenges you have, and learn how your new friend would deal with your challenges. This can be stimulating and invigorating and may take you in directions you never imagined.

Question Conventional Wisdom

The commercial culture in which we live has a powerful interest in getting you to operate (which is to say consume) in standard, conventional ways without thinking about what you are doing. Be prepared to challenge this from time to time at work and in your personal life. Use caution, however, when it comes to challenging work-related issues unless you're sure you have a much better solution than what is being done now.

> "The commercial culture in which we live has a powerful interest in getting you to operate ... in standard, conventional ways without thinking about what you are doing."

Ask the Function Question

Our culture brainwashes us into accepting standard solutions to common problems, and it is easy to get sucked into a rut of conventional thinking. As a behavior analyst, you are trained to ask about the function of behavior. Now ask that same question about larger issues you have to deal with every day. Parents ask for help with their children's disruptive, out of control behavior; shouldn't we ask ourselves how this happened and why? Schools use suspension as a punisher, but it doesn't appear to work. What are the alternatives? We usually get paid by the hour, but shouldn't we really be paid instead by the results we achieve?

Challenge the Status Quo

The test of aggressive curiosity is whether you can make a difference with a new idea you've discovered. Digging deeper for an answer, saying "no" to a conventional proposal, bringing a new person or new perspective to the table, and pushing it hard will make you and others uncomfortable, but it is probably a good sign that you are on to something. You are respecting your business client if you learn everything you can about the history of the company, have memorized the table of organization, and have read the annual reports for the past 5 years. In a recent consultation, we were asked to figure out what it would take to get computer hardware sales associates to push software add-ons. It didn't take long to determine that the company had long-standing incentives for hardware sales but none for software. It had never occurred to the company managers that it was this absence of an incentive that was the issue, not stubbornness on the part of the associates. The company managers were locked into a theory of behavior that we were able to challenge, and we could see the furrowed brows of "Why didn't we think of that?" on their faces.

SUMMARY

As a behavior analyst, you have a unique perspective on the world and much to learn about how everyone else approaches the very same problems that you do. You will be doing yourself and your client a favor if you work to enhance your own curiosity, allow yourself to think broadly about common community and cultural problems, and acquaint yourself with the views of others.

FOR FURTHER READING

Bailey, C. (2000). *God, Dr. Buzzard, and the Bolito Man*. New York: Anchor Books.

Bar-Eli, M., Azar, O., Ritav, I., Keidar-Levin, V., & Schein, G. (2005). Actim bias among elite soccer goal keepers: The case of penalty kicks. *Journal of Economic Psychology, 28*(5), 606–621.

BBC news, Sept. 24, 2008. http://news.bbc.co.uk/2/hi/asia–pacific/7633989.stm

de Bono, E. (1992). *Serious creativity: Using the power of lateral thinking to create new ideas.* New York: HarperCollins.

Gelb, M. J. (2004). *How to think like Leonardo da Vinci: Seven steps to genius every day.* New York: Bantam Dell.

Gelb, M. J., & Caldicott, S. M. (2007). *Innovate like Edison: The success system of America's greatest inventor.* New York: Penguin Group.

Gershenfeld, N. (1999). *When things start to think.* New York: Henry Holt.

Newspaper circulation rising globally, down in U.S. (2008, June 2). *USA Today.* Retrieved from www.usatoday.com/news/world/2008-06-02-newspaper_N.htm.

O'Brien, F., & Azrin, N. H. (1970). Behavioral engineering: Control of posture by informational feedback. *Journal of Applied Behavior Analysis, 3,* 235–240.

Penn, M. J., & Zalesne, E. K. (2007). *Microtrends: The small forces behind tomorrow's big changes.* New York: Grand Central Publishing.

Sutherland, A. (2006). *Kicked, bitten, and scratched: Life and lessons at the world's premier school for exotic animal trainers.* New York: Penguin.

Sutherland, A. (2008). *What Shamu taught me about life, love, and marriage.* New York: Random House.

Conclusions: Action Plan for Behavior Analysts

Now more than ever, it is important for us, as behavior analysts, to review and upgrade our professional skills. We are in competition with many other professions that place a much greater emphasis on the expert business, personal communications, and creative problem-solving skills deemed essential in the marketplace today. Consumers in desperate need of services often completely overlook behavior analysis because they are simply not aware that we exist. Sometimes we do not stand up under close scrutiny by discriminating clients.

Presenting a very confident professional demeanor at meetings, conferences, and one-on-one sessions with clients gives us the opportunity to explain our approach to treatment. Behavior analysis skills alone are not enough, however. We have to be willing to learn from others in business, communications, and consulting. As it turns out, these professionals have a great deal to offer because they've been attracting customers and practicing in the public eye for at least 50 years. In the beginning of our field, back in the mid-1960s, we believed if we simply built a better mousetrap then society would come running to adopt our behavior technology and hire us to implement it. It is time to admit that we were mistaken in this assumption. We must sell our approach and put it in competition with all the other approaches, many of which consistently offer a slicker, more sophisticated, and more marketable presentation than what our field offers.

We began in the animal lab, and we still use much of that terminology in our dealings with teachers, parents, and CEOs. We need to step up our game and analyze more carefully the consumers we want to approach. We must understand their needs and appreciate their apprehensiveness at our current view of behavior, which could be quite contrary to what they were brought up to believe. We talk "evidence-based treatment," whereas clients think and talk about just wanting "something that works." We want data; they want trust. We would like to demonstrate experimental control; they simply want their children to improve their grades or their employees to improve customer service.

If you are a graduate student in a program of studies in applied behavior analysis, you might find it useful to take the test in the appendix. You can rate your current skill set and determine where you need to do some extra work to develop your professional skills.

Postscript

One of the most tragic things I know about human nature is that all of us tend to put off living. We are all dreaming of some magical rose garden over the horizon—instead of enjoying the roses that are blooming outside our windows today.

Dale Carnegie

After turning in our manuscript for publication, we celebrated by going on an adventure to find Behavior Cemetery.

APPENDIX: RATE YOUR PROFESSIONAL SKILLS

Instructions: You can use this checklist as a guide for evaluating your professional skills. Go through each item and rate yourself honestly for each one. For those items on which you rate yourself "average" or lower, develop a personal plan for improvement, starting with the materials in the "For Further Reading" section at the end of each chapter.

25 Skills and Strategies	Strong	Good	Average	Weak	Nonexistent
1. Essential Business Skills					
1. Business etiquette					
2. Assertiveness					
3. Leadership					
4. Networking					
5. Public relations					
6. Total competence					
7. Ethics in daily life					
2. Basic Consulting Repertoire					
8. Interpersonal communications					
9. Persuasion and influence					
10. Negotiation and lobbying					
11. Public speaking					
3. Applying Your Behavioral Knowledge					
12. Handling difficult people					
13. Think function					
14. Use shaping effectively					
15. Can you show me that?					
16. Performance management					
4. Vital Work Habits					
17. Time management					
18. Become a trusted professional					
19. Learn to deal behaviorally with stress					
20. Knowing when to seek help					
5. Advanced Consulting Strategies					
21. Critical thinking					
22. Creative problem solving and troubleshooting					
23. Understanding and using power					
24. Training, coaching, and mentoring					
25. Aggressive curiosity					

References and Recommended Readings

Abernathy, W. B. (2000). *Managing without supervising.* Memphis, TN: PerfSys Press.

Allen, D. (2001). *Getting things done: The art of stress-free productivity.* New York: Penguin.

Atkinson, C. (2008). *Beyond bullet points.* Redmond, WA: Microsoft Press.

Ayres, A. J. (1972). *Sensory integration and learning disorders.* Los Angeles: Western Psychological Services.

Baer, D. M., Wolf, M. M., & Risley, T. R. (1968). Some current dimensions of applied behavior analysis. *Journal of Applied Behavior Analysis, 1,* 91–97.

Bailey, C. (2000). *God, Dr. Buzzard, and the Bolito Man.* New York: Anchor Books.

Bailey, J. S., & Burch, M. R. (2005). *Ethics for behavior analysts: A practical guide to the Behavior Analyst Certification Board Guidelines for Responsible Conduct.* Mahwah, NJ: Lawrence Erlbaum Associates.

Bailey, J. S., & Burch, M. R. (2006). *How to think like a behavior analyst: Understanding the science that can change your life.* Mahwah, NJ: Lawrence Erlbaum Associates.

Beckwith, H. (1997). *Selling the invisible: A field guide to modern marketing.* New York: Warner Books.

Behavior Analysis in Practice. Kalamazoo, MI: Association for Behavior Analysis International.

Behavior Analyst Certification Board. Guidelines for Responsible Conduct, August 2004.

Behavioral Interventions. New York: John Wiley & Sons.

Bixler, S., & Dugan, L. S. (2001). *5 steps to professional presence.* Avon, MA: Adams Media.

Blanchard, K., & Johnson, S. (1982). *The one minute manager.* New York: Berkley Books.

Bloch, J. P. (2005). *Handling difficult people.* Avon, MA: Adams Media.

Block, P. (1981). *Flawless consulting: A guide to getting your expertise used.* San Francisco: Jossey-Bass.

Bracey, H. (2002). *Building trust: How to get it! How to keep it!* Taylorsville, GA: HB Artworks.

Bundy, A. C., & Murray, E. A. (2002). Assessing sensory integrative dysfunction. In A. C. Bundy, S. J. Lane, & A. Murray (Eds.), *Sensory integration: Theory and practice* (2nd ed., pp. 3–34). Philadelphia: Davis.

Carlson, R. (1996). *Don't sweat the small stuff—and it's all small stuff.* New York: Hyperion.

Carnegie, D. (1936). *How to win friends and influence people.* New York: Pocket Books.

Carnegie, D. (1962). *The quick and easy way to effective speaking.* New York: Simon & Schuster.

Carnegie, D. (1981). *How to win friends and influence people.* New York: Simon & Schuster.

Covey, S. R. (1989). *The 7 habits of highly effective people.* New York: Free Press.

Daniels, A. (1989). *Performance management.* Tucker, GA: Performance Management Publications.

Daniels, A. C. (2000). *Bringing out the best in people: How to apply the astonishing power of positive reinforcement.* New York: McGraw-Hill.

Daniels, A. C. (2001). *Other people's habits: How to use positive reinforcement to bring out the best in the people around you.* New York: McGraw-Hill.

Daniels, A. C., & Daniels, J. E. (2004). *Performance management: Changing behavior that drives organizational effectiveness.* Atlanta, GA: Performance Management.

Daniels, A. C., & Daniels, J. E. (2005). *Measure of a leader: An actionable formula for legendary leadership.* Atlanta, GA: Performance Management.

Darling, D. C. (2003). *The networking survival guide: Get the success you want by tapping into the people you know.* New York: McGraw-Hill.

Dawson, R. (2001). *Secrets of power negotiating: Inside secrets from a master negotiator.* Franklin Lakes, NJ: Career Press.

de Bono, E. (1992). *Serious creativity: Using the power of lateral thinking to create new ideas.* New York: HarperCollins.

de Bono, E. (2008). *Creativity workout: 62 exercises to unlock your most creative ideas.* Berkeley, CA: Ulysses Press.

Detz, J. (2000). *It's not what you say, it's how you say it.* New York: St. Martin's Griffin.

Doyle, M., & Straus, D. (1976). *How to make meetings work.* New York: Jove Books.

Duarte, N. (2008). *Slide:ology: The art and science of creating great presentations.* Sebastopol, CA: O'Reilly Media.

Foxx, R. M. (1994, Fall). Facilitated communication in Pennsylvania: Scientifically invalid but politically correct. *Dimensions,* 1–9.

Gelb, M. J. (1988). *Present yourself: Transforming fear, knowing your audience, setting the stage, making them remember.* Rolling Hills Estates, CA: Jalmar Press.

Gelb, M. J. (2004). *How to think like Leonardo da Vinci: Seven steps to genius every day.* New York: Bantam Dell.

Gelb, M. J., & Caldicott, S. M. (2007). *Innovate like Edison: The success system of America's greatest inventor.* New York: Penguin Group.

Gershenfeld, N. (1999). *When things start to think.* New York: Henry Holt.

Gladwell, M. (2000). *The tipping point: How little things can make a big difference.* New York: Little, Brown.

Gladwell, M. (2005). *Blink: The power of thinking without thinking.* New York: Little, Brown.

Goldstein, J. J., Martin, S. J., & Cialdini, R. B. (2008). *Yes! 50 scientifically proven ways to be persuasive.* New York: Free Press.

Gottfredson, M., & Schaubert, S. (2008). *The breakthrough imperative: How the best managers get outstanding results.* New York: HarperCollins.

Greene, B. F., Bailey, J. S., & Barber, F. (1981). An analysis and reduction of disruptive behavior on school buses. *Journal of Applied Behavior Analysis, 14*, 177–192.

Greenwood, M. (2006). *How to negotiate like a pro: 41 rules for resolving disputes.* New York: iUniverse.

Gross, T. S. (2004). *Why service stinks ... and exactly what to do about it.* Chicago: Dearborn Trade Publishing.

Gruwell, E., & McCourt, F. (2007). *The gigantic book of teachers' wisdom.* New York: Skyhorse Publishing.

Hall, P. (2007). *The new PR: An insider's guide to changing the face of public relations.* Potomac, MD: Larstan.

Harvard Business School Press. (2004a). *Coaching and mentoring.* Boston: Author.

Harvard Business School Press. (2004b). *Dealing with difficult people.* Boston: Author.

Harvard Business School Press. (2004c). *Face-to-face communications for clarity and impact.* Boston: Author.

Harvard Business School Press. (2004d). *Manager's toolkit: The 13 skills managers need to succeed.* Boston: Author.

Harvard Business School Press. (2005). *Power, influence, and persuasion.* Boston: Author.

Harvard Business School Press. (2006). *Giving feedback: Expert solutions to everyday challenges.* Boston: Author.

Harvard Business School Press. (2007). *Giving presentations.* Boston: Author.

Heinrichs, J. (2007). *Thank you for arguing: What Aristotle, Lincoln, and Homer Simpson can teach us about the art of persuasion.* New York: Random House.

Henderson, J., & Henderson, R. (2007). *There's no such thing as public speaking.* New York: Prentice Hall.

Hoff, R. (1998). *I can see you naked.* Kansas City, MO: Andrews and McMeel.

Hunsaker, P. L., & Alessandra, A. J. (1980). *The art of managing people.* New York: Simon & Schuster.

Jacobson, J. W., Foxx, R. M., & Mulick, J. A. (Eds.). (2005). *Controversial therapies for developmental disabilities.* Mahwah, NJ: Lawrence Erlbaum Associates.

Journal of Applied Behavior Analysis. Bloomington, IN: Society for the Experimental Analysis of Behavior.

Journal of Organizational Behavior Management. Philadelphia, PA: Taylor & Francis Group.

Kelley, R. E. (1981). *Consulting: The complete guide to a profitable career.* New York: Charles Scribner's Sons.

Kirkpatrick, D. L. (1994). *Evaluating training programs: The four levels.* San Francisco: Berrett-Koehler.

Klaus, P. (2007). *The hard truth about soft skills.* New York: HarperCollins.

Kratochwill, T. R., & Bergan, J. R. (1990). *Behavioral consultation in applied settings: An individual guide.* New York: Plenum Press.

Laermer, R. (2003). *Full frontal PR: Building buzz about your business, your product, or you.* Princeton, NJ: Bloomberg Press.

Lakein, A. (1973). *How to get control of your time and your life.* New York: Signet.

Lattal, A. D., & Clark, R. W. (2005). *Ethics at work.* Atlanta, GA: Performance Management.

Lieberman, D. J. (2000). *Get anyone to do anything.* New York: St. Martin's Press.

Lovaas, O. I. (1987). Behavioral treatment and normal educational and intellectual functioning in young autistic children. *Journal of Consulting and Clinical Psychology, 55,* 3–9.

Martin, G., & Pear, J. (2006). *Behavior modification: What it is and how to do it.* New York: Prentice Hall.

Mason, S. A., & Iwata, B. A. (1990). Artifactual effects of sensory-integrative therapy on self-injurious behavior. *Journal of Applied Behavior Analysis, 23,* 361–370.

Maurice, C., Green, G., & Luce, S. (Eds.) (1996). *Behavioral intervention for young children with autism.* Austin, TX: Pro–Ed Publisher.

McIntyre, M. G. (2005). *Secrets to winning at office politics: How to achieve your goals and increase your influence at work.* New York: St. Martin's Griffin.

McQuain, J. (1996). *Power language: Getting the most out of your words.* New York: Houghton Mifflin.

Michalko, M. (2001). *Cracking creativity: The secrets of creative genius.* Berkeley, CA: Ten Speed Press.

Miltenberger, R. (2003). *Behavior modification: Principles and procedures* (3rd ed.). Belmont, CA: Thompson.

Nalebuff, B., & Ayres, I. (2003). *Why not? How to use everyday ingenuity to solve problems big and small.* Boston: Harvard Business School Press.

National institute for occupational safety and health. (1999) www.cdc.gov/niosh/stresswk.html. Publication No. 99–101.

Newspaper circulation rising globally, down in U.S. (2008, June 2). *USA Today.* Retrieved from www.usatoday.com/news/world/2008-06-02-newspaper_N.htm

Normand, M. T., & Bailey, J. S. (2006). The effects of celeration lines on accurate data analysis. *Behavior Modification, 30*, 295–314.

O'Brien, F., & Azrin, N. H. (1970). Behavioral engineering: Control of posture by informational feedback. *Journal of Applied Behavior Analysis, 3*, 235–240.

Oliver, D. (2004). *How to negotiate effectively.* Philadelphia: Kogan Page.

O'Neill, M. B. (2000). *Executive coaching with backbone and heart: A systems approach to engaging leaders with their challenges.* San Francisco: Jossey-Bass.

Pachter, B., & Magee, S. (2000). *The power of positive confrontation.* New York: Marlowe.

Pande, P., Neuman, R., & Cavanagh, R. (2000). *The Six Sigma way: How GE, Motorola, and other top companies are honing their performance.* New York: McGraw-Hill.

Parkinson, J. R. (1997). *How to get people to do things your way.* Lincolnwood, IL: NTC Business Books.

Paul, R. W., & Elder, L. (2002). *Critical thinking: Tools for taking charge of your professional and personal life.* Upper Saddle River, NJ: Pearson Education.

Penn, M. J., & Zalesne, E. K. (2007). *Microtrends: The small forces behind tomorrow's big changes.* New York: Grand Central Publishing.

Reid, D. H., & Parsons, M. B. (2002). *Working with staff to overcome challenging behavior among people who have severe disabilities.* Morganton, NC: Habilitative Management Consultants.

Reynolds, G. (2008). *Presentationzen: Simple ideas on presentation design and delivery.* Berkeley, CA: New Riders.

Risley, T. R., & Hart, B. (1968). Developing correspondence between the non-verbal and verbal behavior of preschool children. *Journal of Applied Behavior Analysis, 1*(4), 267–281.

Roberts, W. (1987). *Leadership secrets of Attila the Hun.* New York: Warner Books.

Robinson, D. G., & Robinson, J. C. (1995). *Performance consulting: Moving beyond training.* San Francisco: Berrett-Koehler.

Rummler, G. A., & Brache, A. P. (1995). *Improving performance: How to manage the white space on the organizational chart.* San Francisco: Jossey-Bass.

Sagen, C., & Druyan, A. (1997). *The demon-haunted world: Science as a candle in the dark.* New York: Ballantine Books.

Silber, L. (1999). *Career management for the creative person.* New York: Three Rivers Press.

Skinner, B. F. (1953). *Science and human behavior.* New York: Macmillan.

Skinner, B. F. (1957). *Verbal behavior.* Englewood Cliffs, NJ: Prentice Hall.

Smith, T. C. (1984). *Making successful presentations: A self-teaching guide.* New York: John Wiley & Sons.

Smith, T., Mruzek, D. W., & Mozingo, D. (2005). Sensory integrative therapy. In J. W. Jacobson, R. M. Foxx, & J. A. Mulick (Eds.), *Controversial therapies for developmental disabilities.* Mahwah, NJ: Lawrence Erlbaum Associates.

Stone, W. (2005). *Reach for your dreams graduate: Rechange your life with true and courageous stories of individuals who would not accept defeat.* Lakeland, FL: White Stone Books.

Sturmey, P. (2008). Best practice methods in staff training. In J. K. Luiselli, D. C. Russo, W. P. Christian, & S. M. Wilczynski (Eds.), *Effective practices for children with autism.* New York: Oxford University Press.

Sutherland, A. (2008). *What Shamu taught me about life, love, and marriage.* New York: Random House.

Vandeveer, R. C., & Menefee, M. L. (2006). *Human behavior in organizations.* Upper Saddle River, NJ: Prentice Hall.

Walton, M. (1986). *The deming management method.* New York: Dodd, Mead & Company.

Weismann, J. (2006). *Presenting to win: The art of telling your story.* Upper Saddle River, NJ: Pearson Education.

Whitmore, J. (2005). *Business class: Etiquette essentials for success at work.* New York: St. Martin's Press.

Wilder, D. A., Chen, L., Atwell, J., Pritchard, J., & Weinstein, P. (2006). Brief functional analysis and treatment of tantrums associated with transitions in preschool children. *Journal of Applied Behavior Analysis, 39,* 103–107.

Wooden, J., & Jamison, S. (2005). *Wooden on leadership.* New York: McGraw-Hill.

Zechmeister, E. B., & Johnson, J. E. (1992). *Critical thinking: A functional approach.* Pacific Grove, CA: Brooks/Cole.

Zemke, R., & Woods, J. A. (1999). *Best practices in customer service.* New York: HRD Press Amherst.

Index